Saving America's Amazon

Saving America's Amazon

The Threat to Our Nation's Most Biodiverse River System

Text and Photos by

BEN RAINES

Foreword by E. O. Wilson

NewSouth Books | Montgomery

For Jasper and Shannon

NewSouth Books
105 S. Court Street
Montgomery, AL 36104

© 2020 by Ben Raines
All rights reserved under International and Pan-American
Copyright Conventions. Published in the United States by
NewSouth Books, Montgomery, Alabama.

Library of Congress Cataloging-in-Publication Data
Names: Raines, Ben, author.
Title: Saving America's Amazon / Ben Raines ;
foreword by E.O. Wilson.
Description: Montgomery, AL : NewSouth Books, [2018]

| Includes bibliographical references and index.
Identifiers: LCCN 2018050527 (print) | LCCN
2018051539 (ebook) | ISBN 9781588383495 (Ebook) |
ISBN 9781588383389 (hardcover)
Subjects: LCSH: Wildlife conservation--Alabama.
| Aquatic biodiversity conservation--United
States--Congresses.
Classification: LCC QL84.22.A2 (ebook) | LCC QL84.22.
A2 R35 2018 (print) |
   DDC 333.95/409761--dc23
LC record available at https://lccn.loc.gov/2018050527

Design by Randall Williams
2nd Printing   Printed in South Korea

Page 1: Swamp butterweed blooms profusely in the bottomlands as soon as winter floods recede.
Overleaf: A mixed flock of white ibis, tricolor herons, and egrets nest in a rookery on tiny Lady Island each spring.
Above: Thousands of no-name creeks crisscross the cypress bottoms surrounding 200,000 acres of Delta forests.

Unless otherwise indicated, all photographs are by the author.

Contents

The ultimate predator—a lynx spider with its ensnared prey, a bumblebee. At this moment, the spider is sucking the juices from the bee's head. They always seem to start with the head. Lynx spiders can spit venom, which is potent enough to humans to cause temporary blindness and irritate the skin.

Foreword

Edward O. Wilson

No natural paradise has ever endured more savage treatment for so long yet better survived than Alabama's Mobile-Tensaw Delta. I have loved it in dreams and reality. As a teenage boy during the early 1940s, I often rode my bicycle out of Mobile across a bridge rimming the Delta's northern edge, then turned to enter the causeway that separates its southern border from Mobile Bay. Thence to complete a ten-mile journey all the way to the end at Spanish Fort and Blakeley (where one of my Confederate ancestors had fought in the final battle of the Civil War). In recent years I've followed by automobile most of the same route, and taken trips into the forest on flat-bottomed boats.

Within the almost impenetrable depths of the great forest little has changed. It is still possible to see and feel the core of the primeval forest that existed thousands of years ago. The American Amazon, as Ben Raines points out in this valuable account, earns its byname for a reason beyond its resemblance to a tropical rainforest. The section of the Gulf Coast in which the Delta is centered, together with the deep ravines of the Red Hills directly to the north, harbors the richest per hectare fauna and flora found in North America. The opportunities for original research and education in environmental science are immense.

The Mobile-Tensaw Delta has been tough and resilient as one of America's premier natural areas. It is there to enjoy and explore, to inspire and teach. Yet, as Raines also makes clear, much of the past two centuries of agricultural and industrial growth along the central Gulf Coast depend on its source rivers and natural harbor of Mobile Bay. As a consequence the Delta forest and lacework of waterways through it have deteriorated in part. May future generations appreciate and learn to conserve its timeless treasures.

Alabama native and two-time Pulitzer Prize winner E. O. Wilson is the preeminent biologist of his time. He popularized biodiversity while discovering dozens of new species. He created the field of sociobiology, and he is considered the world's leading authority on ants. In his later years, Wilson has championed the diversity of his native state, particularly the Mobile River Basin, which he says is, from a scientific perspective, "as unexplored as Borneo."

Ants pick over
the remains of
a dragonfly that
fell for the sticky
red tendrils of the
sundew, which
can ensnare bugs,
lizards, and even
small birds. Prey
is attracted by
both scent and
what appear to
be glistening
dew drops. Once
an animal has
wandered into
the gooey trap,
its fate is sealed,
as it slowly
starves to death.
As the carcass
decomposes,
its rotting body
nourishes the plant.

Preface

This book is based on a series of articles published in the *Mobile Press-Register*, the *Birmingham News*, and the *Huntsville Times* while I was the executive director of the Weeks Bay Foundation environmental group. Those articles, in turn, were based on the documentary *America's Amazon*, which I wrote and produced with Lynn Rabren and Mary Riser. Many of the interviews in this book were first conducted for either the film or the newspaper series.

Saving America's Amazon is the culmination of twenty years spent covering Alabama and the Gulf Coast as a reporter focused primarily on the environment. I spent those years documenting and fighting the horrors we heap on the birds, fish, trees, wetlands, and waters in the state of Alabama, which wreaks greater harm on our wild places than any other state.

To bear witness to this devastation is too much for many. My longtime editor at the *Press-Register*, a botanist of the highest caliber and winner of a bushel of environmental journalism awards, packed it in years ago and took a job with the Nature Conservancy. He told me that it had become too hard to try to protect Mother Nature with pens and public opinion.

His new job required him to hunt for the last, best wild places in Alabama and ask rich people to buy them and protect them. Those places will be preserved, surely a good thing, but what of the rest of the state? What of the places where we live,

swim, and get our drinking water? Will a few protected acres of pristine pitcher plant bogs and old growth longleaf pine—acres scattered amongst and surrounded by tree farms and Superfund sites and subdivisions like so many islands in a barren sea—be enough? The answer, of course, is no. Not in Alabama. Not in America, where we are not nearly so good at protecting air and water, people and wild animals from the caustic effects of industry and development as most of us would like to believe.

For years I lashed out broadly from a now nearly defunct newspaper in Mobile, a sleepy Gulf Coast port town set on a swampy bay in the rainiest spot in America. My words sent corporate vice presidents scurrying back to Houston and New York and other monied places with scuttled plans for billion dollar factories on the shores of my bay. I proved that oil platforms in the Gulf of Mexico qualified as Superfund sites based on the contamination in the creatures and sand surrounding them. I found a town where the roads were paved with mercury waste, literally billed as a gift to the populace from the local chlorine factory. I was spied on in my home by the largest corporation in the world, and I captured two animals that were thought to be extinct. I wrote about a woman whose drinking water had so much methane in it you could light it on fire as it ran from the tap, and I won several of the biggest environmental journalism awards in the country.

In the center of this photo, a two-foot alligator, with only his eye visible, is poised in a clear spot next to a thick mat of vegetation. He's waiting for a frog to come hopping by on the vegetation. Small gators make their living this way, while larger specimens hunt hogs, deer, turtles and dogs. Either way, the green body and muted colors are purpose-made to help this predator ambush its prey.

Writing stories about bad people doing bad things to the environment felt like my calling. It made me feel the way I imagine deeply religious people feel. I loved that job. I poured myself into it.

And then corporate masters in New Jersey gutted the three largest papers in Alabama, slashing staff and cutting publication to just three days a week. In one fell swoop, millions of people were robbed of vital information about what was going on in their own backyards. Dozens upon dozens of reporters were fired in each of the state's largest cities. Gone were city council reporters, business reporters, school reporters, and on and on. Suddenly, no one was watching the hen house. Especially when it came to the environment. There can be no question that the polluting industries noticed.

The demise of our once robust newspapers is especially worrisome in Alabama, where officials spend less to protect the environment than in any other state, despite having one of the richest assemblages of plants and animals seen on Earth. Here, as all across the nation, newspapers have been the environment's best friend, a pipeline for the public to drop a dime on the man dumping a barrel of crap in a river or a power company shoveling PCBs into a ravine. With those voices being slowly strangled in our fractured media landscape, protecting Mother Nature has become much, much harder.

And so—this book you hold in your hands.

This is my attempt to make you fall in love, to care, to do something about this incredibly rich and rare place we call Alabama, before it falls to ruin, one species, one acre, one stream at a time.

The indigo bunting is the bluest blue you can imagine.

ACKNOWLEDGMENTS

I couldn't have written this book without the time and knowledge shared with me by dozens of scientists, naturalists, fishing guides and friends. For two decades, I have been privileged to tromp through forests, crawl around pitcher plant bogs, and travel thousands of river and ocean miles with people who know more than I do about the natural world that surrounds us. I consider it an education beyond anything one could hope for at the finest university. It has defined the course of my professional life. To anyone who has ever gifted me a day afield or afloat with them, please accept my humblest thanks. I will name a few who deserve special acknowledgement for their help with this book.

Bill Finch, my old newspaper editor, was the first to take me into a pitcher plant bog and served as my first Delta guide. His affection and excitement for Alabama's unique landscapes were infectious. The scientists with the Dauphin Island Sea Lab, Auburn University and the University of South Alabama allowed me to spend hundreds of days in their labs, on their research vessels, at field test sites, or archaeological digs.

Especially generous with their time were Bob Shipp, John Dindo, John Valentine, Sean Powers, Judy Stout, Ken Heck, George Crozier,, Ruth Carmichael, Steve Szedlmayer, Mimi Fearn, Greg Waselkov, John Mcreadie, Crystal Hightower, and Marcus Drymon. Other scientists whose work was directly influential include Andy DePaola, Monty Graham, Roger Clay, Steve Heath, Peter Tuttle, Kristine DeLong, Martin Becker, Harry Maisch, Harriet Perry, Ed Cake, Gena Todia, J. J. Apodaca, and Marco Kaltofen.

Through hundreds of hours on boats together, fishing guides Bobby Abruscato, Richard Rutland, and Jeff Dute helped unlock the mysteries of Mobile Bay and the Gulf of Mexico the way only people who live their lives on the water could. The help and knowledge of Brian Jones, curator of the Sea Lab's Estuarium, has been invaluable as he willingly accompanied me everywhere from the bottom of the Gulf of Mexico to the tiniest streams. He helped find many of the creatures seen in these pages.

Finally, my thanks to my mother, Susan Raines, who gave me my first camera and taught me how to take pictures, and my father, Howell Raines, who made sure we always had a boat.

Saving America's Amazon

The Mobile River system, which drains most of Alabama, is far and away the most diverse river network in North America. More than one-third of all known freshwater fish species in the U.S., Canada, and Mexico are native to Alabama's rivers and creeks. Colorado, twice the size of Alabama, has 55 species of fish, compared to more than 450 species in Alabama, with new species discovered regularly.

Overleaf: From its beginning as an ephemeral stream near the town of Megargel to its confluence with the Alabama, Little River is a mere 26.1 miles long. Despite its short run, it is home to more species of fish than are found within the borders of many states.

TENNESSEE

MISSISSIPPI

ALABAMA

GEORGIA

Mobile Bay

Gulf of Mexico

Map courtesy of Mobile Bay National Estuary Program

1

Saving America's Amazon

Hidden away in the heart of Dixie, one of the nation's greatest wildernesses is being destroyed, bit by bit, in a silent massacre.

You won't find armies of hippies chaining themselves to trees to protect this place, or national environmental groups using pictures of it to sign up new members, because almost no one even knows it exists. And yet, here it is—the Mobile River Basin, the culmination of the richest river complex in North America, and one of the richest in the world in terms of the sheer number of species and types of habitat.

The basin's river system stretches from the top of Alabama to the bottom, drains parts of four states, and encompasses hundreds of thousands of acres of forest, from Appalachian hardwood stands to haunted cypress swamps. Its waterways are great and small—from giant rivers to tiny creeks you could step across without getting your feet wet. A dedicated band of locals know it for the incredible hunting and fishing it affords.

But almost no one knows it for its greatest distinction. That's a shame, for this is America's Amazon.

In the ascendant view among scientists, this river network and its surrounding forests represent the true cradle of America's biodiversity. There are more species of oaks on a single hillside on the banks of the Alabama River than you can find anywhere else in the world. Thanks to the Mobile River Basin, the state of Alabama is home to more species of freshwater fish, mussels, snails, turtles and crawfish than any other state. And the contest isn't even close.

For instance, Alabama is home to ninety-seven crawfish species. Louisiana, famous the world over for boiled crawfish, has just thirty-two species; California, three times the size of Alabama, has but nine. There are four hundred and fifty species of freshwater fish in the state, or about one-third of all species known in the entire nation. The system's turtle population is even more singular. When it comes to turtles, the system's estuary, the Mobile-Tensaw Delta, has eighteen species, more than any other river delta anywhere in the world. More than the Amazon. More than the Mekong. More than any other river system on Earth.

Unlike most of the nation's great river systems, the Mobile Basin—along with its wetlands, floodplain forests, and estuary—has survived with its biological community mostly intact. That is due in large measure to an odd combination of benign neglect and the mixed blessing of being located in the heart of Alabama.

Tragically, this incredible system now sits on the cusp of decline, facing death by a thousand cuts, just as the scientific community has begun to appreciate its riches. Habitat destruction, development, and lax enforcement of environmental regulations conspire to take an increasing toll, making the area a global hot spot for extinctions, particularly of aquatic creatures. In fact, nearly half of all extinctions in the continental United States since the 1800s have occurred among creatures that lived in the Mobile River Basin, according to records maintained by Endangered Species International and the U.S. Fish and Wildlife Service.

Saving this native bounty—which ought to be preserved as one of the nation's greatest treasures—will take a sea change in state regulations, public attitudes, and priorities. The first step on that path is to revel in the existence of this wilderness, richer by almost every measure than any other place in the nation, yet scarcely known even to the people who live closest to it.

This notion of Alabama as biologically rich is a recent phenomenon, first confirmed in NatureServe's 2002 compendium on American diversity, *States of the Union*, in which Alabama ranked first in aquatic diversity and among the top five states for the overall number of species of plants and animals. It is worth noting that the four other states in the top five for the most species are all more than twice as large as Alabama by area and have less than half as many species per square mile.

Shielded from wide-scale scientific inquiry in the twentieth century simply because Alabama was a long way from the nation's most prominent research universities, few biologists explored the state. That is changing now, and new species are discovered regularly in both the state's flora and fauna.

In a recent interview, Alabama native E. O. Wilson, one of the world's most-famous scientists, summed up the scientific awakening occurring in regards to the Mobile Basin as it has occurred in his own life. He spoke after an expedition into Mobile River Basin forests where he captured several ant species believed to be new to science. His personal scientific journey began here, where, as a boy, he caught snakes and bugs on the edges of the Mobile-Tensaw Delta.

"I've traveled the world exploring its most exotic places. And now I find that the most magical of all was right here where I started out all those years ago," Wilson said. He has described the mission of his later years as bringing scientific attention to this area, and in so doing, save it.

"Can you imagine, more turtles than anywhere else in the world. It's just marvelous. People have no idea the wonders that will come from that system. It's as unexplored as Borneo. When we talk about the advances in medicine and biochemistry and other fields that might be waiting to be discovered in that place, it's really just remarkable," Wilson said.

Yet Alabama goes out of its way to hide its glory. Despite

The Mobile-Tensaw Delta has the most turtle species—17—of any of the world's river deltas. This little guy, the stinkpot turtle (*Sternotherus odoratus*), lives up to its name by releasing a foul odor from glands along the edge of its shell. Predators are likely to search for a more palatable meal.

hundreds of native wildflower species, including showboats like the Red Hills azalea, American lotus, and wild delphiniums, the state flower is a camellia, native to China, Japan, and Korea. The Alabama Wildlife Foundation gives out an annual "Governor's Conservation Award." When the Weeks Bay Foundation, where I used to work, won the award, it featured a statue of a mountain goat, native only in the northernmost Rocky Mountains, mostly on the Canadian side.

Sadly, Alabama's incredible firmament of natural wonders was understood centuries ago. It was celebrated on both sides of the Atlantic beginning in the 1700s. The state's natural wonders played a role in the educations of visionary scientists in the fields of geology, botany, and biology. Yet that legacy was largely forgotten by the end of the nineteenth century, when the state became more famous for its coal deposits.

In some measure, Alabama's natural heritage has been obscured by a century's worth of bad press that framed the state in terms of what we've done to it, and in it, rather than what has always been present in the landscape beneath our feet. Civil rights

The grass pink orchid (above), or any of the other 53 orchid species in the state, or the Red Hills azalea (above left) could surely be Alabama's state flower rather than the non-native Asian camellia. But tellingly, Alabama officials feature a mountain goat, native to the northern Rocky Mountains, on the Governor's Conservation Achievement Award.

Right, a blue bottle fly eats coyote dung full of deer hair in a Delta floodplain forest. Hundreds of species of flies have been identified in the Delta, including dozens new to science. Below, green fly orchids live in the tops of live oaks.

protests, steel mills, cotton fields, and football—that's the Alabama living in the mind of the public. But there is another Alabama, a wild Alabama, that ranks among our country's rarest and most precious assets.

In that hidden Alabama, there are more species of flesh-eating pitcher plants than can be found anywhere else on Earth. Nine new species of bottle flies were discovered in a single day in 2012. There are fifty-four species of orchid, from the tiny green fly orchid—an arboreal species living in the tops of ancient live oaks—to the white fringed orchid, whose delicate flowers glow like Chinese lanterns in the crepuscular hours before dawn.

Over the years I have interviewed Dr. John McCreadie, an entomologist at the University of South Alabama, a number of times. He is in the midst of a major multi-decade effort to catalogue the Mobile-Tensaw Delta's insect species, and he authored a chapter in a National Park Service study of the Delta's history and biology. So far, McCreadie's team has identified eight hundred species, with thousands more awaiting identification by specialists. Asked to tell me what he could about the Delta's insect population, he said, "There's so much we don't know."

McCreadie's team has caught four hundred species of flies in the Delta—about twenty-five entirely new to science. The preliminary work suggests the Delta is a "hot spot" for insect diversity. Contrast that with the state of knowledge just a few years ago, when McCreadie invited a noted expert on a group of flies known as long-legged metallic flies to come sample the Delta forests. The expert declined on the grounds that Alabama was only known to have nine species of long-legged flies.

Since then, McCreadie has discovered ninety species of long-legged flies in the Delta, including fourteen new species. "We are really just beginning to understand what's out there," McCreadie said.

Alabama's wild places have long benefited from a predominantly rural economy and sparse development, as the state missed out on the great industrial and technological growth spurts seen elsewhere in the nation since the 1950s. But today,

Alabama is poised on the edge of a period of prodigious expansion. Thanks to hefty tax credits, business-friendly labor laws, and lax environmental regulation, the state has become a manufacturing hub—with Mercedes, Hyundai, and Honda all making cars at new assembly plants. An Australian company now builds battleships for the Pentagon in Alabama, and Europe's Airbus began assembling its largest jetliner on the shores of Mobile Bay in 2017. A newly completed steel mill built on hundreds of acres in the heart of the Mobile-Tensaw Delta is the largest and most expensive manufacturing facility constructed in the U.S. in the past quarter century. Subdivisions and cities are spreading quickly across acreage long devoted to timber, agriculture, or kudzu.

With that growth comes the likelihood of accelerated destruction for what scientists have only begun to realize is one of the most diverse ecologies outside of the Amazonian rain forests.

The decline is already beginning to show. Mobile Bay, the catch basin for this river system—which drains parts of Mississippi, Tennessee, and Georgia—now has annual dead zones like the well-publicized ones in the Gulf of Mexico, due to excess nitrogen and other nutrients flowing from fertilization of farms and suburban lawns. Meanwhile, the rivers of the Mobile Basin have been dammed and channelized and bear the scars of industrial pollution, with swamps that rank among the most polluted in the nation for DDT, mercury, and other contaminants. Our once-clear rivers are now often a muddy brown, a testament to the amount of runoff coming from farms, urban areas, and subdivision construction sites.

The fall of 2016 proved just how precarious things have become, as numerous major rivers and streams ran completely dry, in some cases for the first time ever. While this in fact was during a drought, the real culprits were the various industries—paper mills, coal mines, steel mills, golf courses—that in the absence of government oversight or regulation were allowed to suck up the last drops from the rivers. Despite having more miles of rivers and creeks than any other state, Alabama has no official rules governing the amount of water that industries may draw from them. Instead, companies along riverbanks are guaranteed rights to the water no matter what environmental havoc may occur. On the

Newborn praying mantises looking for a first meal stalk a white-topped pitcher plant.

Tallapoosa River, fish rotted in fetid pools after its flow stopped. On the Coosa, the river disappeared for miles as irrigation wells for golf courses and cotton fields sucked the groundwater dry, draining the river from below. The dried-up Cahaba only began flowing again downstream of the sewage treatment plant for the city of Trussville. Incredibly, the entire flow of one of the state's major rivers was from treated water from a sewage plant. And then consider this: the Cahaba River is home to one hundred and fifty species of fish, more species than you find in the entire state of California. Imagine, roughly one-sixth of all the freshwater fish species known in the United States live in a single Alabama river that is just one hundred and ninety-four miles long.

Alabama's environmental protection laws are simply not up to the task of safeguarding the state's vast treasures. Alabama ranks last in the nation in spending to protect the environment. The state's environmental agency has been zero-funded by the legislature for years. The Alabama Department of Environmental Management is now fighting a petition in federal court that seeks to compel the U.S. Environmental Protection Agency to take over the administration of the Clean Water Act in the state due to thousands of non-prosecuted alleged violations.

AGAINST THIS BACKDROP, THE moment has arrived when Alabama must take up and celebrate its incredible natural heritage, or sit by as it is slowly and forever diminished. The stakes are real. For all the appreciation the state deserves for being home to more

Done in by a beetle—the bare limbs and brown leaves represent redbay trees killed by an invading Chinese beetle no larger than a grain of rice. Redbay were once one of the five most common trees in our swamps. They've been almost completely wiped out in less than a decade, an extinction occurring before our eyes. The beetle is also causing the extinction of the spicebush swallowtail butterfly.

species of fish, mussels, snails, and crawfish, Alabama is also at the top of another list—of aquatic extinctions.

Each year, a few more species are lost, and another piece of the nation's richest biological treasure is gone forever. The extinction crisis extends to terrestrial species as well, as seen in records kept by the U.S. Fish and Wildlife Service, Endangered Species International, and NatureServe. The rate of extinctions in Alabama is roughly double that seen anywhere else in the continental United States.

Alabama has lost ninety species, followed by California at fifty-three species, and California is three times larger. On top of the ninety species confirmed as extinct in Alabama, scientists think an additional ninety species may have slipped into extinction, and the U.S. Fish and Wildlife Service suggests that another hundred species should be protected under the Endangered Species Act. Alabama has more extinctions than the combined total of the surrounding states of Louisiana (seven), Mississippi (eleven), Florida (twenty-three), Georgia (twenty-six), and Tennessee (twenty-two), according to the Nature-Serve analysis.

Of course, part of the reason there have been so many extinctions in Alabama is because there were so many rare species here to begin with. The task now is to turn the tide before the state loses the crown for highest aquatic diversity in the nation and is left only with the crown for the most creatures lost forever to extinction.

A hidden houseboat floats in a secret lagoon in the most diverse river system in the country, home to creatures and plants yet to be discovered. If, that is, we find them before they are lost to development and pollution.

Thirty-five-million-year-old sand dollars spill out of the banks of the Alabama River near Claiborne Bluff, a testament to when this spot and most of Alabama was 300 feet beneath the primordial Gulf of Mexico.

2

Meet America's Amazon—Not Always Fresh, But Never Frozen

The Mobile River Basin and the thousands of creeks and rivers feeding it together form the largest inland delta system in the United States, encompassing all of the tributaries of the Cahaba, the Coosa, the Tallapoosa, the Black Warrior, the Alabama, and the Tombigbee. It is second only to the Mississippi in how much water it dumps into the Gulf of Mexico, and the flow from the Mobile system is more than thirty times the combined flow of all the other river systems east of the Mississippi.

The state as a whole boasts seventy-seven thousand miles of flowing rivers and creeks, scattered from the Appalachian foothills surrounding Birmingham and Huntsville at the top of the state to the Gulf seashore 390 miles south. In between are mountains, ancient rock formations, grasslands, prairies, sheer cliffs, hundreds of miles of brackish shoreline, fifty miles of Gulf beaches, hundreds of miles of brackish bay shorelines,

and an exceptionally complex geological make up. Mix it all together, stir in one of the nation's rainiest and mildest climates, and out comes Alabama, which biologist and author Scot Duncan affectionately calls "a Goldilocks state, not too hot and not too cold," in his excellent book, *Southern Wonder*.

In fact, ice, or the lack of it, is perhaps the most important factor behind the great abundance of life seen in the Mobile River system, compared to western rivers such as the Columbia, or the eastern systems such as the Potomac or the Hudson. The ice in question is not the ice that forms in modern winters in those colder, Northern rivers but the ice that blanketed much of North America for centuries at a time during the cyclical ice ages that have visited the continent over the last one hundred million years, and especially the last two million years.

Going back in time just ten thousand years provides a window into the changes wrought by successive ice ages. Back then, both the Columbia River Valley and the Potomac's broad floodplain, where Washington, D.C., sits today, suffered under tundra-like conditions, as did most of the United States. In places, the ice sheets were nearly a mile thick. Even places that weren't covered by glaciers were so close to the ice that their climates were too cold to support most life. Similar conditions have occurred repeatedly throughout geologic time, roughly every forty thousand to one hundred thousand years, according to the ancient climactic records. The Earth is still thawing out from the most recent ice age, but you can see the effects of the last several ice periods in the fossil record, which shows ancient plants and animals that were lost across North America during the big freeze.

But not in Alabama, which was saved from the ravages of the Pleistocene glaciers due to its lower latitude. While most of the United States was frozen solid during the last several ice ages, Alabama was not. It was colder here than it is today, to be sure, and windier, but Alabama's hills and valleys and coast remained warm and hospitable. When plant and animal life beneath or along the edges of the vast sheets of ice was snuffed out at the beginning of the last ice age about eighteen thousand years ago, wiping away eons worth of evolutionary change all across North America, it was as if someone hit a reset button and forced the continent's forests to start over from scratch. Most American forests are actually less than ten thousand years old, which, in the grand scheme of things, is not much time for evolution to occur.

In Alabama, life survived, cloistered away in deep valleys in the rocky northern part of the state, or in the warm open meadows of the longleaf pine forests in the Black Belt and coastal plain.

"Oh, this is an amazing place. This is the kind of place where you find things left over from before the glaciers came and changed the climate," said Mark Deyrup, with the Archbold Biological Station, standing in a stream deep in a valley in the Red Hills. Deyrup and a team of scientists working with E. O. Wilson are part of a concerted scientific effort to document new species in Alabama. The group believes the Red Hills region in the center of the state represents a biological oasis unlike anything remaining in North America, with potentially hundreds of unknown species, ranging from ants and spiders to wasps, salamanders, and plants.

Comprised of a band of bluffs and valleys a few miles wide and fifty miles long, the Red Hills stretch in a skinny line across five Alabama counties. The hills themselves are actually ancient coral reefs. The valleys are spaces between the old reefs, where time and rain eroded the softer sediments, washing them downstream to the Gulf. The area is home to dozens of plant species considered more typical of the Appalachian Mountains, hundreds of miles north. Geologically, the hills are the oldest part of the Appalachians, the first bones of a range that runs north to the Canadian border. The forests on these hills are likely the richest in the United States. On a single hillside you can find twenty species of oak trees, making this area the global center for oak diversity. Compare that to the entire expanse of the Great Smoky Mountains National Park, which brags about having a mere dozen oak species within its borders, or to the entire state of California.

Below, the dense roots of tens of thousands of acres of Roseau cane (locally, *Row- Zo*) bind the Delta mud in place during floods. Reaching to ten feet tall, these canes are last year's crop, gone to seed. Down closer to the ground, new green shoots are emerging.

The slopes here are steep, too steep to scuttle up without grabbing hold of a tree root or one of the hunks of sandy, crumbly pieces of ancient coral and marl poking through the fallen leaves on the forest floor. Even with a handhold, the moist leaves underfoot slip away in little avalanches, revealing rich black mud underneath. Rushing rivulets, three-inches wide, pop from the hillsides, making for a forest floor dotted with tiny springs. The water bubbles up through the porous limestone of the primeval seafloor underlying the hills. Dripping from mossy overhangs onto chalky, pale yellow siltstone and exposed clay banks the color of dried blood, the water works its way down to creek banks crowded with thickets of azaleas and mountain laurel. Native gingers and trilliums with fanciful names like the bashful wakerobin and the propeller toad shade sprout underfoot. These forests are dark and old, magical and full of possibility.

"The glaciers never got here, but they changed the climate so much that many things just disappeared," Deyrup said. "It either got too cold or too dry for them . . . but they survived in places like this. There is no telling what might still be waiting to be discovered. Anywhere you have a salamander that is in its own genus!"

The Red Hills salamander is what Charles Darwin described as "a living fossil" in his landmark treatise on evolution, *The Origin of Species*. When he wrote of ancient species that "have endured to the present day, from having inhabited a confined area, and from having thus been exposed to less severe competition," he might have been writing a life history of this salamander, which has existed in its present form for perhaps fifty million years.

As a species, it is tens of millions of years older than the rest of the American salamanders. Despite existing for eons, these salamanders are considered an evolutionary dead end, like a duck-billed platypus. Typically, you would expect to find numerous closely related species that evolved from such an ancient line, branching off in a sort of salamander family tree, but none exist. They have no close relatives. This species is tied so closely to the unique habitat it calls home that its adaptations simply won't work anywhere else. Put another way, the Red Hills salamander could exist nowhere else but in these hills, and it never had to try to spread beyond them because these hills have provided a warm and stable home for an eternity.

It is not easy to lay eyes on one of these endangered animals. Though they are about a foot long, they were not discovered until 1960, nearly thirty-five years after the last new American salamander species had been found. They remained hidden for so long because they are fossorial, meaning they live their entire lives underground, in networks of shallow tunnels they carve in the soft mud surrounding the limestone outcroppings in the hills.

The Red Hills salamander is like a duckbill platypus—an evolutinary dead end. Found only in these hills, they live up to 40 years.

To catch one, you must go fishing deep in the woods with no water in sight, armed with hooks so small they might have been crafted by fairies and fishing line as fine as doll's hair. Bait your tiny hook with a cricket, and let the cricket march into a salamander burrow. First, of course, you must find a burrow, in itself no small task. Search through the leaf litter and moss on the hillside for a round hole small enough that you could plug it up with a nickel.

I went salamander fishing with J. J. Apadoca, a biology professor at Warren Wilson College in North Carolina, who has been studying the salamanders for a decade. He is one of the only people on the planet with a license allowing him to capture the rare animals for research. We watched his cricket march into the entrance of a burrow. After it made it about an inch inside, we could see movement. A salamander rushing toward the free meal! The strike, when it came, was fast. The salamander's mouth gaped broadly and swallowed its prey whole. After a brief fight, Apadoca tugged the salamander from its subterranean lair, wiggling on the end of the gossamer line. The moist, purple skin glistened like a wet eggplant in the late afternoon sun. Cradling it in his hands, Apadoca noted that his catch was a female, and pregnant. The eggs were visible through the translucent skin of her swollen belly. Her legs, less than half an inch long, seemed ridiculously short when compared to her total length of eleven inches. But legs aren't much use in the tunnels where the creatures dwell. The Red Hills salamanders are almost never seen outside of their burrows, or in daylight. They prowl the hours between dusk and dawn, sitting just inside the mouths of their caves, waiting for an unsuspecting roly-poly or insect to march by.

The Red Hill salamander is living proof that truly ancient plants and animals persist in Alabama. Most of the plants that live in the eastern United States can still be found somewhere in Alabama, from dozens of species of trees to hundreds of flowering plants, including many species more typical near the Canadian border. Scientists now think of Alabama as a refugium for both plants and creatures, a temperate oasis through the great climactic upheavals. When the glaciers retreated, thousands of plant species spread from the area, slowly moving northward along the spine of the Appalachians, repopulating the eastern forests as temperatures warmed.

"Alabama stayed warm. It was almost like a spring in that regard, where you have stable temperatures year round as the groundwater bubbles up from the earth," said Pat O'Neil, a biologist with the Geological Survey of Alabama, and one of the authors of the *Fishes of Alabama*, one of the first modern scientific explorations of the state's exceptional diversity. "When you have stable environments that have escaped sea level change and the glaciers, then, with time, we're talking millions of years, evolution can proceed forward at pretty rapid rates and you get a lot of speciation. That's what happened in Alabama, particularly in the Mobile River Basin. That area has been that way for millions of years. Combine that with the diverse geology and diverse aquatic habitats we have, and it leads to diverse fauna."

A Southern wood fern unfurls and stretches toward the morning sun on a hillside in the Red Hills.

Opposite: This creek in the Red Hills flows through a series of stair-step pools cut through the limestone marl left behind by an ancient sea. The tiny pools of Alabama streams like this are home to an array of fish found nowhere else on Earth.

The zeuglodon, or basilosaurus in modern parlance, captured the attention of the world when the first specimens were recovered in Alabama in the early 1800s. Described by early scientists as "king lizards," the 60-foot-long creatures were actually among the earliest whales, evolving about 50 million years ago. Their teeth were the size of a grown man's hand, used for attacking sharks and other prehistoric ocean dwellers. Fossil hunters still regularly find basilosaurus teeth in Alabama, and farmers occasional turn up whole skeletons while plowing fields. There was a global trade in the relics for nearly a century, with many specimens looted from Alabama. With even partial skulls fetching tens of thousands of dollars, the fossils are now protected by a law that prohibits their removal from the state without written permission from the governor. This basilosaurus above, on display in the Smithsonian National Museum of Natural History, was collected in Choctaw County, Alabama in 1894. (Courtesy of the Smithsonian Institution)

3

Behold the Zeuglodon!

Alabama, positioned at the southern edge of the continent and next to the modern Gulf of Mexico, has had a wild ride during the last few billion years, and the proof is scattered all across the landscape. In addition to its temperate climate, two other intertwined factors set the stage for the evolutionary explosion that defines Alabama: geology and habitat.

The most notable and recent factor responsible for the state's hodgepodge of geological terrains has been changing sea level, which has varied by hundreds of feet. In fact, going back forty million years to the middle Eocene era, much of Alabama was underwater, submerged under an ancient ocean that covered much more of the earth than the modern seas.

You can see the marine past in the rivers and hills of central Alabama. There, one hundred fifty miles inland, the streambeds are littered with ancient shark teeth, bits of coral, and the remnants of rostrum spikes from swordfish and marlin. The shark species represented include long-extinct giants—such as the forty-five-foot long *Carcharocles auriculatus*, the ancestor of the megaladon—as well as species like the makos and tiger sharks that still swim in the oceans of today.

A fossil hunter armed with a shovel and a colander can collect a thousand teeth in a day—deep blue and polished to a shine by their multi-million-year journey—if he knows where to hunt. Meanwhile, along the Alabama River, the sepia-tinted remains of ancient sand dollars spill from the banks—the delicate and distinctive petal pattern on the surface of each specimen etched as perfectly as when it was alive, some thirty-five million years ago.

Though it is but little known today, Alabama's ancient marine past was celebrated and studied in Europe in the 1800s. In fact, Alabama played a profound but almost unknown role in the development of Charles Darwin's theory of evolution. Our drowned river valleys, particularly along the Alabama River, have been famous among fossil hunters for centuries. In the 1800s, the then-fledgling Smithsonian Institution displayed a massive zeuglodon skeleton found in Alabama, a skeleton that is still on display today at the Smithsonian National Museum of Natural History.

Top, found in the Sepulga River, a tributary of the Conecuh River, this 35-million-year-old tooth belonged to the ancestor of the famous megalodon shark. Above, two hours of sifting riverbed gravel in the Sepulga yielded hundreds of teeth, including from species that are still alive today, such as mako and tiger sharks.

The zeuglodon, or basilosaurus, was a monstrous dinosaur-era whale with razor-sharp teeth the size of a man's hand. The discovery of dozens of these ancient skeletons—each about seventy feet long with mouths large enough to swallow a man whole—in fields in central Alabama was heralded as proof that sea monsters once roamed the world's oceans. More importantly, the finds put Alabama on the map for Victorian-era fossil hounds.

Sir Charles Lyell, an Englishman who is considered the father of modern geology, traveled the state in the 1840s, collecting fossils throughout the Mobile River Basin, as far north as Tuscaloosa and as far south as Mobile. He published a book about this Southern sojourn, *A Second Visit to the United States of North America*. "At this season, January 29th, the thermometer stood at 80 Fahrenheit in the shade, and the air was as balmy as on an English summer day," Lyell noted. A highlight of the trip for Lyell was his arrival at the already famous cliffs of Claiborne Bluff on the Alabama River, where he stayed for several days hunting fossils.

Fossil hunters still flock to the chalky white bluffs today to collect ancient snails, crabs, oysters, and sand dollars, though many of the bluffs are now buried behind the lake made by the Claiborne Dam.

Lyell was able to refine his theories on the age of the earth by drawing comparisons between the processes that formed coal fields and fossils he saw near Birmingham, England, and Birmingham, Alabama, and between the ancient oceans that created the marine deposits at Claiborne and similar deposits in France and Italy.

Based on these observations, Lyell posited that the earth was tens of millions of years old, not six thousand years old as the Church of England maintained at the time. These studies were the basis of his seminal text, *Principles of Modern Geology*, which Charles Darwin carried aboard the HMS *Beagle* on his fateful journey to the Galapagos.

Lyell's influence on Darwin is well documented. During the voyage of the *Beagle*, Darwin wrote in his journal of seeing geologic formations in the Galapagos and South America through "Lyell's eyes." Back in England after the trip, Lyell became Darwin's mentor and close friend and served as a sounding board during the twenty years between Darwin's trip to the Galapagos and the 1859 publication of his landmark treatise on evolution, *On the Origin of Species*.

What Lyell gave to Darwin was an understanding of the element of time—the proof that the earth was old enough for the plants and animals living on it to have had time to evolve over millions of years. It was Lyell who coined the names we use today to describe the prehistoric epochs, such as Pleistocene, Miocene, and Eocene. He attached ages to these eras dating back millions of years, and he showed that by examining the fossils present in an area, it was possible to figure out the age of the rocks found there. Going further, he understood that the species present in fossil assemblages varied with the age of the formations. Newer formations, he discovered, had a higher percentage of species still living today than older deposits, and there were species present in the more recently created formations that were not present in earlier eras.

Inherent in that discovery is proof that new species are created as time passes, although Lyell did not delve into a mechanism for that creation, as Darwin would. Instead, he and other scientists first began to ponder how and why new species appeared in the fossil record as time passed and why there were different species in Alabama than England.

"About 400 species, belonging to the Eocene formation, derived from this classic ground [Claiborne Bluff], have already been named, and they agree, some of them specifically, and a much greater number in their generic forms, with the fossils of the middle division of the deposits of the same age of London and Hampshire," Lyell wrote. Geologists, he continued, "would do well to compare the succession of rocks on the Alabama River with those of the same date in England."

Based on his Alabama visit and other explorations, Lyell noted that different sites around the world, similar in age and with many overlapping species, sometimes had unique species not seen in the other places. From his work, scientists, including Darwin, began to consider the role that unique habitat features, predators, or geologic events such as volcanic eruptions might play in the extinction of some species and the ongoing survival of others.

From there, it was not such a great leap for Darwin to propose his theory of natural selection to explain the diverse species that he encountered in the Galapagos. The animals there had changed with time and circumstance, just as the animals seen in Lyell's fossils had changed in response to the world around them.

The portion of Lyell's book describing his visit to Alabama contains as much social and political analysis as geologic and remains a unique and fascinating historical text.

The first zeuglodon had been discovered just a few years prior to Lyell's visit and caused an international stir, with scientists originally believing the creature to be a giant lizard. By the time of Lyell's visit, the remains of forty such animals had been discovered within a ten-mile radius, all near Clarkesville. Most were found by farmers turning their fields with plows. The giant vertebrae were popular as chairs in homes in the area and were sometimes used as foundation supports for farmhouses. Lyell noted that the discoveries were a popular topic of conversation among locals.

A cockle fossil with an array of delicate fingers protruding from its shell in a limestone bed at Claiborne Bluff.

Right: A tiny fossil of the Nassarius genus of marine snails, found about 120 miles inland. Its descendants still scavenge in the Gulf of Mexico.
Below: Paleontologist Martin Becker points to an Eocene era shell revealed by erosion along Claiborne Bluff on the Alabama River.

"Many were curious to learn my opinion as to the kind of animal to which the huge vertebrae, against which their plows sometimes strike, may have belonged," Lyell wrote. "The magnitude, indeed, and solidity of these relics of the colossal zeuglodon, are such as might well excite the astonishment of the most indifferent."

Lyell reported that "the remains of the zeuglodon have been also found by Mr. Hale in this cliff [at Claiborne]." Mr. Hale was the Rev. C. S. Hale of Mobile, who was fascinated by fossils and published several scientific texts on Alabama fossils in the 1850s. This was a provocative pastime for a preacher in an era when Christian thoughts on science were defined by Bishop Usher's decree that the Bible showed the earth to be six thousand years old.

In relating his attempts to find a zeuglodon skeleton to examine, Lyell offered a succinct social commentary on Alabama, circa 1845:

> I witnessed no maltreatment of slaves in this state, but drunkenness prevails to such a degree among their owners, that I cannot doubt that the power they exercise must often be fearfully abused. In the morning the proprietor of the house where I lodged was intoxicated, yet taking fresh drams when I left him, and evidently thinking me somewhat unpolite when I declined to join him. In the afternoon, when I inquired at the house of the German settler, whether I could see some fossil bones discovered on his plantation, I was told that he was not at home; in fact, that he had not returned the night before, and was supposed to be lying somewhere drunk in the woods, his wife having set out in search of him in one direction, and his sister in another.

Lyell went on to write with amazement that the practice of dueling continued in the South, despite being outlawed, and to offer his observation that "instead of the ignorant wonder, very commonly expressed in out-of-the-way districts of England, France, or Italy, at travelers who devote money and time to a search for fossil bones and shells, each [Alabama] planter seemed to vie with another in his anxiety to give me information in regard to the precise spots where organic remains had been discovered."

Lyell's interest in the state prompted more attention, which in turn led to the appointment of the first state geologist, who catalogued coal fields and fossil deposits with equal vigor.

"In taking up the study of the Coastal Plain of Alabama, attention was naturally turned to this part of it, the river banks. How well this promise has been fulfilled," reads a report on the state's fossil deposits from 1887 by E. A. Smith, state geologist of Alabama. "We have in this section of Alabama the most complete and varied series of Eocene and Cretaceous strata known in the United States."

The strata Smith referred to include limestone outcroppings left behind when seafloors were uncovered by receding oceans thirty-five million years ago, as well as the more ancient Cretaceous deposits dating back 140 million years, to the tail end of the age of dinosaurs. There are even older marine deposits as well, from five hundred million years ago. Those primeval, chalky seafloor deposits, and the minerals contained within them, have played a central role in the development of the unique suites of plants and aquatic creatures occupying the wetlands and streams of the state. The soils of the Black Belt region, for instance, are defined by the fossil-rich geologic formation they sit upon, known as the Selma Chalk, itself an ancient seabed.

Going back even further, the state also boasts rock outcroppings dating back more than a billion years, when the earth was rent by massive changes and continents caromed around the oceans like pinballs, crashing into each other with such force as to create mountain ranges like the Appalachians, which reach their southern terminus in Alabama.

The serrations along the edges of a 40-million-year-old fossilized tooth remain razor sharp.

Shaped by its ancient geologic history, the state today has more than five dozen distinct habitat types, including coastal dunes and marshes, sawgrass meadows similar to the Florida Everglades, grasslands home to perhaps more species of grasses than can be found in all of the Midwestern prairies, floodplain jungle as thick as anything in the Amazon, and the rocky and mountainous lands in the northernmost corners, which are home to the most diverse hardwood forests in the United States.

The loop current enters the Gulf of Mexico carrying warm tropical water from the Caribbean. It spins up toward Louisiana, and then cuts east toward Alabama and Florida. The current acts as a conveyor belt for delivering rain to the Alabama coast throughout the year, with the moisture generated by the warm water seeding the thunderstorms that drop an average of 70 inches of rain on south Alabama each year. Passing through the Florida Straits, the current feeds into the Gulf Stream current that keeps Europe warm. (Google Earth, Data SIO, NOAA, U.S. Navy, NGA, GEBCO Image Landsat / Copernicus Image © 2020 TerraMetrics)

4

A Desert But for the Rain

By all rights, the entire state of Alabama ought to be a desert, and the incredible array of rivers and streams in the state ought to be as dry as a Death Valley gulch. Study a globe, tracing a line around the world between the thirtieth and thirty-fifth parallels, a band that neatly encompasses all of Alabama, north to south, and you find most of the world's hottest deserts.

At the southern extreme of the Mobile Basin, the city of Mobile shares its latitude with Cairo and the vast Sahara desert. Meanwhile, Montgomery and Birmingham share latitudes with Phoenix, Arizona, and Albuquerque, New Mexico, respectively, along with the Mojave Desert. But far from being a dry and forbidding expanse, Alabama ranks as one of the top three wettest states on average and is home to the single rainiest city in the nation. Seattle gets a meager fifty-five inches of rain a year on average, compared to more than seventy inches a year in Mobile and the area surrounding Mobile Bay.

The Gulf of Mexico acts as a conveyor belt delivering warm, moist air from the region around the equator to the southern United States. There is also a water current in the Gulf, known as the Loop Current, that spins up from the Caribbean, brushes along the tip of Louisiana's boot, and then turns east, headed for the Florida Straits, between the Florida Keys and Cuba. That current, along with atmospheric forces, guides warm masses of tropical air north across the Gulf. When the air masses, laden with tropical moisture, collide with the fronts and weather systems moving across the continent, they begin to drop their load as rain. One of the first places that moist air hits land is Alabama.

All that rain means a steady supply of water for the seventy-seven thousand miles of rivers and creeks in the state. And all that water, coupled with the same amount of sunshine that bakes the desert surrounding Egypt's Pyramids, makes for a place that bursts at the seams with life.

A new study of the Alabama Delta region funded by the National Park Service concludes: "This combination of moisture, light, and heat sustains high rates of primary productivity in terrestrial and aquatic systems, and sustains globally significant

Above: Area of detail.
A satellite image of the Mobile-Tensaw Delta and upper Mobile Bay. The Delta gathers water that falls in parts of Alabama, Mississippi, Tennessee, and Georgia. The heavy flow creates an ever-changing network of rivers and creeks that collectively form the third-largest river delta system in the nation. The great swamp, situated between urban areas to the east and west that are home to half a million people, serves as a major migration corridor for millions of birds annually, as well as bears, panthers, and manatees. (Google Earth, Data SIO, NOAA, U.S. Navy, NGA, GEBCO Image Landsat / Copernicus Image © 2020 TerraMetrics)

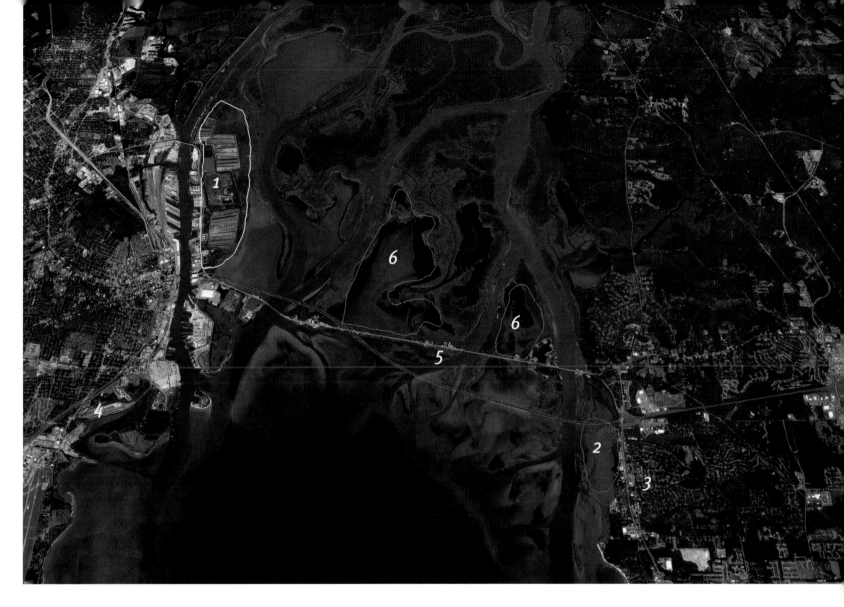

Destroying a Delta —

For 100 years, development in Mobile has meant destruction for the lower Delta. The (1) area outlined in yellow was once known as Polecat Bay. From the 1940s until the 1970s, Alcoa Aluminum and the U.S. Army Corps of Engineers filled 900 acres of open water and seagrass meadows with dredge spoil and the toxic waste generated by Alcoa. The (2) area outlined in blue was known as D'Olive Bay. The entire bay, once 9 feet deep and about 320 acres, filled with mud during the construction of the (3) Lake Forest subdivision in the 1970s–'80s. The (4) area outlined in red, Garrows Bend, was one of the most productive marshes in Mobile Bay until 2014, when the State of Alabama chose to destroy the marsh and build a rail yard to facilitate the unloading of container ships. Meanwhile, the (5) U.S. 98 Causeway was built atop an earthen dike across the head of the bay in the 1920s. The (6) areas outlined in pink were deep bays prior to the construction of the roadway and are now almost completely filled thanks to sediment trapped by the roadway.

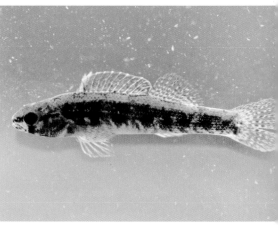

The Evolution of Darter Species

Alabama's 77 darter species inhabit fast-flowing streams statewide. Mature darters are about three inches long. Their varied coloration makes a powerful case for Darwin's theory of evolution. *Top row*, from left: Trispot darter, Redline darter, Watercress darter. *Second row*: Tombigbee darter, Redspot darter, Greenbreast darter. *Third row*: Tallapoosa darter, Gold Stripe darter. *Bottom row*: Greenside darter, Slackwater darter, Coldwater darter. (Images courtesy of Bernie Kuhajda, Tennessee Aquarium)

floral and faunal diversity. The only temperate region on Earth with comparable biological productivity and biodiversity is Southeast Asia."

The Mobile Basin begins in the rocky region at the top of the state. There, hidden in the folds of the old mountains, raindrops and spring-fed rivulets gather between sheer granite walls and hillsides rich with iron ore. These smallest mountain streams come together, ankle deep and gin clear, gaining steam as they begin a tumbling journey downhill to the faraway sea.

Biological rarities begin in these streams, home to creatures such as the vermillion darter, one of the most fantastically colored—and critically endangered—freshwater fish to be found anywhere in the world. At just two inches long fully grown, it makes its living hiding in the eddies behind stones in riffle sections of its home stream, darting out to grab the nearly microscopic nymphs of tiny stonefly and caddis species carried on the current. Its streamlined body sports tall, turquoise and brick-red sails for dorsal fins and broad fans at the pectorals

designed to produce the power needed to propel the tiny fish through its fast-water home.

Alabama has seventy-seven darter species that inhabit the smallest capillary streams—with fanciful names that make a sort of poetry when strung together: lipstick, lollipop, greenbreast, and redbelly; blueside, brighteye, dusky, and naked; slenderhead, harlequin, holiday, and Johnny. Over time, separated for eons by physical barriers, such as mountains and waterfalls, that prevented the mixing of populations, darters living in streams just a few miles from each other evolved into a dizzying array of unique species, a pattern repeated among the other aquatic fauna in the state. Some species, like the vermillion, are known to live only in a single stream, an existence as close to the edge of extinction as one can imagine.

The cane cutter rabbit is an accomplished swimmer, frequently crossing creeks and sloughs in the summer. But cold winter floods are more than the rabbit can handle. The Delta is transformed into a rushing river 13 miles wide, with no dry land to swim to. Mammals and reptiles are forced to find any high ground, be it a tree, vine, or even a floating log to sit for weeks or even months to wait out the floods.

Alabama's tiniest streams, tens of thousands of them, flow together to form rocky rivers full of cataracts and the richest populations of aquatic snails and freshwater mussel species anywhere on the planet. The state is home to 182 mussel species, representing 60 percent of the species in the entire nation. Of those, forty-eight species are considered critically imperiled.

The Cahaba River alone, which has its beginnings in the hills near Birmingham, was historically home to fifty species of mussels, or about one-fifth of all known species in North America. Biologists say that due to extinctions, the river is now home to only twenty-seven species of mussels.

The stream banks of the Cahaba, shaded by hardwood oak and hickory forests, are lined with an Appalachian mix of mountain laurel, rhododendron, and native azaleas. But the banks are also brushed with colors not seen at points north, such as a deep purple larkspur—scientific name *Delphinium alabamicum*—with flowers the size of your thumb. Or in the stream itself, tall green spears topped with the spider-like Cahaba lilies, which wave in cotton-white patches above the riffles. It makes for a landscape that is at once familiar to anyone who has frequented the rivers of the eastern United States and yet strangely unique.

Moving south, we come to the Fall Line, visible on satellite images of the state as a curving scar that arcs through Opelika to Montgomery then curls north through Tuscaloosa and Tuscumbia. A legacy of those ancient changes in sea level, the Fall Line also serves as a demarcation between the rocky top half of the state and the coastal plain to the south. The Black Belt region begins below the fall, with dark, fertile soils, underlain by a hard layer of chalk left behind by the ancient seas. The Selma Chalk, millions of years old, defines the Black Belt and controls the plants and animals that can live on its prairies even today.

For the big rivers—the Alabama, the Cahaba, the Black Warrior, and the Tombigbee—the Fall Line represents a significant change. Below the fall, no longer bound by rocky valleys, the rivers spread out into broad alluvial valleys, meandering through forested floodplains that are subject to significant annual flooding for months at a time during the winter and early spring, like the Amazon or the Nile. The rivers in this part of the state are big and deep, carrying vast amounts of water collected from the hills of parts of Tennessee, Georgia, Mississippi, and Alabama. An entirely different collection of species—shiners, chubs, suckers, and even saltwater species from the Gulf of Mexico—are commonly found in these river reaches.

Meanwhile, the annual flooding fuels an incredibly fecund ecosystem from this point south, as rains upstate gather the decaying remains of the previous summer's bounty and spread them downstream. The sediment deposits that spill out when the rivers spread below the fall line are rich with nutrients that feed prolific spring growth in one of the most diverse temperate jungles on the planet.

As the Black Belt soils peter out, the rivers of the Mobile Basin meet the Mobile-Tensaw Delta, the largest inland river delta in the United States. While the Mobile River Basin ranks as only the fourth largest system in the country by flow, it is the undisputed champion in terms of species diversity. Located just two hundred miles to the west, the mighty Mississippi River Delta, which drains 41 percent of the nation, looks woefully depauperate by comparison.

This Delta forest is dry land most of the year. The floods begin between December and February and can last through March. The ground is about six feet underwater in the above image, and the current is powerful — boaters beware! The floodwaters gradually recede, leaving the tupelo and sweetgums with their toes in the water for months at a time.

Inset: You can always recognize a swamp that stays wet most of the time by the swollen bottoms of the trees.

Deep in the Delta, waterways like this seldom see human visitors because fallen trees block boat traffic. The only way in involves a chainsaw.

5

Where the Rivers Meet the Sea

E. O. Wilson describes the importance of the Mobile-Tensaw Delta on a global scale: "It is arguably the biologically richest place. The Delta floodplain forest and swamp, and the area immediately around it, including the Red Hills to the north, has more species of plants and animals than any comparable area anywhere in North America, the United States and Canada. . . . it is a place yet completely unexplored, sort of like the upper Amazon."

The Delta proper is a vast jungle wilderness where dozens of river channels braid together and twist apart, creating hundreds of islands large and small. Those islands are populated with numerous creatures capable of killing a grown man, including bears, alligators, bull sharks, bobcats, feral hogs, and five species of venomous snakes. Add to that three hundred bird species, seventy-three amphibians, ninety-three reptiles, and tens of thousands of species of insects.

But even with all its diversity, more than anything, the Delta is remarkable and essential for being the place where all the water running downhill from the rest of the state meets the sea.

The drama that unfolds as the rivers hurl their mighty plume of fresh water against the expanse of the Gulf of Mexico is ancient and primal. It is also changeable, with the river system winning the battle part of the year and the Gulf winning part of the year. The push and pull of salt versus fresh water remakes and reenergizes the Delta landscape, with the cast of creatures and plants living off the bounty changing on an almost weekly basis.

Each spring, the annual flooding turns dry land wet as hundreds of thousands of acres of bottomland forest in the Delta floodplain disappear under a wall of water ten feet deep for three to five months. Imagine an area fourteen miles wide and forty-five miles long—that's about how much land disappears under water during the spring floods, when the Delta is given over to the aquatic world, and terrestrial creatures, such as snakes, squirrels, rabbits, and bears, are either up in the trees or have managed to scramble up the bluffs that define the Delta's edges.

The woods are spooky during the flood, and treacherous.

The purple and lavender of the iris fields startles the eye as the flowers suddenly trumpet their presence. It is only when the iris bloom that their significant role in the swamp ecosystem becomes apparent. They are one of the most important plants holding the mud in place.

46

Currents pushing ten miles an hour sweep through the thick forests. Landmarks and trails vanish along with every bit of land. The Delta is silent now, its summer sounds—the buzz of hundreds of thousands of insects per acre and the chatter of an army of birds—replaced by the rushing swirl of water against tree and the deep bellow of spawning alligators. A boater who loses power in the flooded woods quickly finds himself in a dangerous situation, pinned against the cypress and gum trees by the muddy brown water. The biggest danger is turning over or falling out of your vessel. Do that and drowning is almost guaranteed. If you manage not to be pulled under by the deceptively strong currents, a slower death awaits as hypothermia takes hold, for it is a sure bet that no one will ever find something as inconsequential as a lone human clinging to one of the millions of trees in five hundred square miles of flooded forest.

When the flood is on, the plume of muddy river water takes over Mobile Bay, turning the salty, twenty-mile long estuary into a freshwater extension of the Delta. The brown river water flowing down the bay into the Gulf of Mexico is visible from outer space, just like the Amazon's plume. Scientific monitoring conducted by the Dauphin Island Sea Lab has traced the freshwater influence fifty miles offshore.

But the tide slowly turns. The spring rains slow. The sun takes hold along with the summer drought. More water evaporates. As the rest of the nation heats up, the air currents that have pushed moist tropical air—and the rain it carries—from the equator to the north are less powerful. Less rain means less flow, and the rivers fall back within their banks. But the floods have done their work, sweeping the floor clean in the bottomland forests and leaving behind rich alluvial deposits that fire the startling spring and summer growth in the low country.

The first flush comes in purple, painted across the huge open meadows of the lowest reaches of the Delta. More purple than blue, our native iris is nonetheless known as the blue flag. For a slim window every spring, tens of thousands of the flowers glow along the creeks surrounding hidden bays and bayous and in between the knees of the cypress. The loose, drooping petals flutter in the breeze, sprouting from big, sword-like leaves rooted in the swamp mud. The roots below are one of the key things holding much of the Delta in place, though their importance is only apparent when the iris bloom and it becomes clear just how many there are.

Some are the deep purple of the sky on a moonless night, while others seem made of watercolors, cast in fading lavenders or the washed-out blue of an old pair of Levis. The rarest are a clear and icy hue so faint it is almost gray. No matter the color of the flower, splashed down the middle of each petal is a flash of gold that glows like starlight. It is astonishing to see such a showpiece spearing up from nothing, from bare mud.

The flowers are so lovely, and their arrival so brief and unexpected, it is no wonder that they have inspired poets and revolutionaries for generations. It was an iris that inspired the famous French fleur-de-lis, which became the symbol of France and later New Orleans.

The swirl of colors worn by the purple gallinule—closely related to coots—is mesmerizing. Duck hunters are allowed to shoot these gorgeous birds.

After months underwater as nymphs, hundreds of thousands of mayflies hatch within hours, taking to the skies for a 24-hour whirl of mating and egg-laying. The adults then die.

And it was a flag emblazoned with an iris that Joan of Arc carried into battle against the English.

It is easy enough to see great numbers of iris from the bow of a boat cruising through the lower portion of the Mobile-Tensaw Delta. Still, my favorite spot for iris-gazing demands a little more work. It is also in the Delta, and the only way to get to it is by boat. But to see these flowers, you must get off the boat and into the mud. Think of it as a sort of toll.

The price is fairly high. Many I've taken to this most secret garden turn around before we get to the payoff. I understand why. With every step, you sink well past your ankles in muck. And about every tenth step, you are likely going in up to your knees. And then there are the snakes. About half of them are the harmless brown water snake. But the other half are poisonous cottonmouth moccasins, which have a tendency to curl up in the nook between two branches of the tree you are about to grab.

The brave few who have made it all the way in have witnessed what surely ranks as one of the loveliest spots in all of Alabama. Imagine a shaded fairy bower, perhaps two acres in size, where every piece of ground is covered by a sea of purple iris flowers, waving on a gentle breeze. It was just such a place, I imagine,

that inspired Henry Wadsworth Longfellow when he wrote his poem about the iris, "Flower-de-Luce," in March of 1866. The poem ends this way:

> Thou art the Muse, who far from crowded cities
> Hauntest the sylvan streams,
> Playing on pipes of reed the artless ditties
> That come to us as dreams.
> O flower-de-luce, bloom on, and let the river
> Linger to kiss they feet!
> O flower of song, bloom on and
> Make forever
> The world more fair and sweet.

As the iris start to fade, the snow-white Delta lilies emerge. For a time, both are blooming, the lilies stark against a greening sea of foliage that is still spiked a thousand times over by the purple iris. Standing in one of these soggy-bottomed spring meadows during this bloom is akin to the dizzying feeling of staring into a kaleidoscope.

By June, the iris and lilies have burned out, and the Delta's plume of fresh water is but a shadow of its former self, no longer strong enough even to push its way into Mobile Bay.

This is when the buzzing begins. On summer afternoons, the very air in the Delta throbs. From dragonflies to gargantuan black grasshoppers called lubbers, the sounds of Delta summer are all about the birds and the bees. A massive, Delta-wide blitz of gluttony and sex unfolds as the creatures of the realm fornicate and gorge themselves during the salad days and nights of June.

And then it is time for another flower, more spectacular still. Rounding a bend in the Tensaw River in these early days of summer, the eye is immediately drawn toward the far bank,

where, even at great distance, a broad expanse of gigantic golden flowers can be seen spreading away from the water's edge. Behold the American lotus!

A thousand lemony yellow flowers the size of salad plates bob above a lush carpet of green lotus pads that stretches for a quarter mile along the bank. If the word pad conjures an image of the classic lily pad floating on the water's surface, think again, for the leaves of the lotus are almost as remarkable as the magnificent flowers.

The lotus pads stand nearly two feet above the water on stout stalks, which support the leaves—great green bowls of foliage that look velvety to the touch and repel water better than any fabric man has yet invented. The flowers rise a foot above the pads, making for a striking display.

The naturalist William Bartram was taken with the lotus when he encountered the flowers in the Delta in the late 1700s. "The surface of the water is overspread with its round floating leaves, whilst these are shadowed by a forest of umbrageous leaves with gay flowers, waving to and fro on flexible stems, three or four feet high," Bartram wrote. "These fine flowers are double as a rose, and when expanded are seven or

A red-wing blackbird sings from atop the stalk of a lotus blossom it just ate. The birds devour the seed pods at the center of every lotus.

Lotus stands line the riverbanks in the lower Delta, bursting into bloom in July and August.

eight inches in diameter and of a lively lemon color."

These are the plants of "lotus-eater" fame, the New World version of the lotus eaten by the sailors in Homer's "Odyssey." The only real difference between this American species and those found in Asia and Africa is the color of the flower. Ours are yellow, while theirs are pink.

Regardless of their color, the sight of so many on the riverbank casts the same spell anywhere on the globe. "They started at once, and went about among the Lotus-eaters," wrote Homer, describing what happened to three of his sailors. The lotus, he wrote, "was so delicious that those who ate of it left off caring about home, and did not even want to go back and say what had happened to them, but were for staying and munching lotus with the Lotus-eaters without thinking further of their return."

As it gets hotter and drier, the salt takes over. The Gulf of Mexico turns Mobile Bay saltier by the day. By August, the brackish estuary is dominated by clear salty water, and saltwater species, including Spanish mackerel, tarpon, and sharks, are spread throughout the entire length of Mobile Bay. What's more, the Delta itself becomes salty as Gulf tides push inland eighty miles, carrying blue crabs, shrimp, flounder, redfish, and speckled trout into the deepest corners of the freshwater swamp, where the marine creatures feed on the bounty generated by the spring floods.

Along the swamp's leading edges, freshwater plants, such as the lotus and wild rice, temporarily die back as salt-loving species gain a foothold, with salt marsh grasses momentarily colonizing what was, just weeks before, a freshwater swamp.

These flowers, closely related to the Cahaba lilies of north Alabama, are known as Delta lilies in south Alabama. They are native across the South and perfume the night air with an otherworldly scent.

Clockwise from top right: A global glider clings to an arrow arum leaf popping up in the lotus. The glider is the insect with the largest known migration, sometimes crossing oceans. | An alligator glides through a layer of duckweed floating on the water's surface. Duckweed, which flourishes in the quiet backwaters out of reach of the river current, is the smallest known flowering plant. | A tiny largemouth bass, caught during a survey by state biologists. The Delta's bass switch to a diet consisting primarily of saltwater creatures in late summer, consuming the shrimp and crabs growing in the grass beds. | The first thing the lubbers eat is the flower. It appears to be quite a delicacy. Only after they've eaten all of the flowers on a stem will they work their way down to the seeds and stalks.

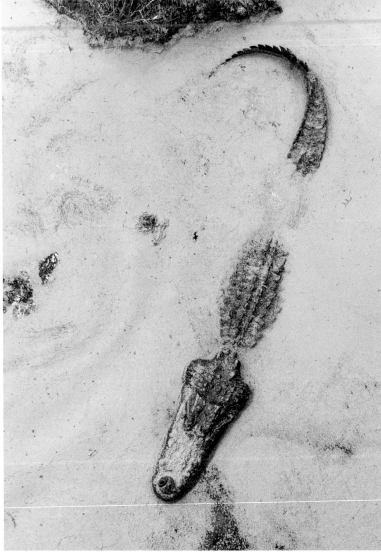

"In the spring and late winter, really it is a freshwater world—largemouth bass, carp, these kinds of things. Then in the summer, when the freshwater inflow recedes, it switches over to a more marine community. That's when the sharks start showing up, that's when mangrove snapper start showing up, bluefish, flounder," said John Valentine, director of the Dauphin Island Sea Lab. "What's unique here, you have this one system where you have these apex predators from different worlds, salt and fresh, colliding. You'll have bull sharks and alligators, and both are feeding on bluegill or menhaden or whatever they find in the swamps, but they are competing with each other as top predators. Then you have the manatees in the Delta, and bears, and feral hogs. You have all these large animals occupying this one place. That doesn't happen other places."

In 2009 biologists observed a bull shark swimming around the base of the Claiborne Dam, ninety miles inland on the Alabama River. Meanwhile, just a few miles from the foot of the same dam, hunters caught the largest alligator ever captured in the United States, a fifteen-foot, thousand-pound monster. In its belly was a 115-pound deer that had been eaten whole.

As for the manatees, the giant sea cows migrate annually from wintering grounds in central Florida to summer in the Mobile River Basin. Analysis of their fecal deposits shows they too take advantage of the Delta's riches. While the manatee diet in most of their range consists primarily of aquatic plants, the hundred-plus manatees that spend their summers in Alabama are enjoying an unprecedented smorgasbord.

"One of the most surprising things we see with the manatees in the Delta is that they are reaching out of the water and eating plants growing on land, cattails and things, and they are eating large amounts of acorns. They are clearly finding areas with oak trees growing over the water and eating the acorns that collect there," said Allan Aven, a manatee biologist with the Dauphin Island Sea Lab.

The Delta forests producing those acorns are a wonder all their own. Surrounding the edges of the vast floodplain, and growing in the dry places in the middle of it, the forests represent one of the continent's most unique assemblages of plants. Water and time are the gardeners here, sculpting and fertilizing the land with floods and forever reworking places where plants might find a foothold.

In an old oxbow known as Byrnes Lake, along the eastern edge of the Delta, about ten miles from Mobile Bay, there exists a forest almost schizophrenic in nature, which is representative of the Delta as a whole. It is at once Appalachian and tropical, modern and prehistoric, familiar and otherworldly.

In an area perhaps the size of a tennis court, you can find flavors of all the eastern forests in North America, and you can find plants that grow nowhere else but this one spot. The result is a forest that could exist only in this place, where these rivers have washed against these riverside bluffs for hundreds of thousands of years, where one of the richest gardens to be found on Earth has had time to grow.

Imagine for a moment you are standing in the center of a single acre of this swamp forest. Steep hillsides butt right up to the edge of the water here. Their moist and shady flanks are carpeted with a half-dozen species of trilliums and groves of wild anise, an evergreen shrub so aromatic you smell it long before you see it. The anise flowers are dainty and oxblood red, and look like sea anemones. Rare orchids grow high in the trees above your head. At your feet, a stand of waxy and silver Indian pipe flowers pokes through the leaf litter, the one plant in the entire forest with no chlorophyll. Their pale, almost translucent stems and flowers are the swamp's ghosts, living a saprophytic existence.

A trio of juvenile yellow crowned night herons wait for their mother's return. When they develop adult plumage, they will be covered in gray feathers, with pronounced yellow mohawks of feathers down the centers of their heads. Their enormous eyes hint at how they make their living: prowling the shallows in the twilight hours.

Meanwhile, the understory surrounding you is shaded by trees you might find on a hillside in West Virginia or New Jersey—American chestnuts, maples, oaks, mountain laurel, and beech. At the water's edge in our acre, long slender fingers of golden club, which have been flowering in the American swamps for more than one hundred million years, lend a primeval flavor, while saw palmetto and beautyberries ripen beneath head-high sumac. Passionflowers and potato beans twist their vines up tree trunks and over fallen logs. Buttonbush and red bays fight for real estate in the thick black goo left by vanishing floodwaters. And, just a few feet away, cypress trees grow as islands unto themselves out in the black water of the

swamp. Come spring, the bases of the cypress trees are fringed with a petticoat of the daintiest yellow flowers—the native St. John's wort, which often colonizes the tiny islands that gather around the cypress trunks.

There are many more species present in this acre, but just in those named above are compounds with medicinal applications for depression, spasms and convulsions, prostate problems, poisoning, pain, liver issues, eye problems, skin disorders, bacterial infections, and cancer.

There, in the very plants, is the living treasure of the Delta and the entire Mobile Basin, said E. O. Wilson, referring to the hidden potential contained in its biodiversity. He likened the entire river system to the Amazon basin, where researchers are now pitted against loggers in a race to unlock medical discoveries before the forests are destroyed.

"We know that many of these species and organisms here, if we study each one deeply, are going to reveal secrets and tell us about life and medicine, biochemistry, and so on," Wilson said. "Most people have no idea how little explored a system like this is, and what great yields to human welfare and medicine a system like this can give. This is its great importance."

Left: A barred owl surveys the swamp floor looking for crayfish, mice, snakes, squirrels, and other prey. *Above and right:* The prothonotary warbler's plumage screams "look at me!" to find a mate. Living in the deep greens of our springtime swamps, the brilliant yellow feathers do catch the eye. The birds stand about 2.5 inches tall and rely on their splashy coloration for finding each other in the vast forest. They are seldom seen more than a few hundred feet from a creek, lake, or, most often, a swamp. They nest in hollows in dead trees, and it is hard to find a place with more dead trees than an Alabama swamp. They migrate as far as Colombia in South America, and return to Alabama every year.

Above: A tiny tree frog clings to a burnt stick left over from a spring fire. Next to him, a rose pogonia orchid begins to open in the morning dew. Right: A white violet is the first green thing out of the ground after a fall burn at the Nature Conservancy's Splinter Hill bog.

6

The Bog—A Real-Life Little Shop of Horrors

The rarest of all the Mobile Basin ecosystems, home to a number of species that grow nowhere else in the world, are the pitcher plant bogs, scattered from the top of the state to within a few hundred feet of the salty Gulf coastline. Wetland habitats by nature, the bogs occur on the hillsides surrounding the rivers, where thousands of tiny springs weep through the ground, or along the wet edges surrounding the creeks and swamps, or in broad meadows covering hundreds of acres in the rainiest parts of the state, where the ground stays moist year round.

The bogs could only have evolved along with the majestic longleaf pine, whose light and airy canopy allows plenty of sunlight to reach the ground, and whose pine straw serves as a perfect fuel source for quick, hot forest fires. Those fires, sparked by one of the thousands of lightning strikes visited on the state annually, are the key ingredient required to maintain pitcher plant bogs. In fact, the bogs evolved largely as a response to fire. This region of the Gulf Coast is home to more lightning strikes than anywhere else on the continent, with Alabama alone recording more than eight hundred thousand strikes a year. In effect, fire is nature's gardener here, burning away young oaks and woody shrubs that would otherwise shade out the rich diversity of life that springs up in a newly burned longleaf pine savanna. Many of the plants in the bog, from toothache grass to wiregrass to the large and showy grass pink orchid, only flower after fires, timing their reproduction with the burst of nutrients released by fire.

Walking through Splinter Hill, a five-hundred-acre bog complex jointly owned by the state and the Nature Conservancy on the western edge of the Mobile-Tensaw Delta, is like taking a step back in time, to Alabama as it was for the thousands of years before modern farming and land use. In that Alabama,

The dew-covered webs seen in meadows on early mornings are spun by the scarlet sheetweaver, which preys on gnats and other tiny bugs.

Hill, atop a narrow ridge spine, broad, park-like meadows spread before you, falling off onto gentle hillsides that stretch as far as you can see. The trees, almost exclusively long-leaf pines close to one hundred feet tall, are spaced ten and twenty feet apart. A carpet of knee-high grasses runs off into the distance. Splashes of red, yellow, blue, pink, purple, and green spear up above the grass, testament to the hundreds of species of flowering plants that surround you. There are pale pink orchids and golden honeycomb sunflowers. Hibiscus and milkweed crowd around huckleberry bushes loaded with fruit that will ripen by summer. And butterflies—monarchs and frittilaries, cabbage whites and hairstreaks—dash from flower to flower, suckling at a lady's tresses orchid, then at the pale lilac flower of a thread dew. Daisies, meadow beauties, bog buttons, candy root, wild peas, and asters are all coming out now. The ground is aglow with their blossoms.

The most conspicuous of all are the pitcher plants, carnivorous wonders that have man-

the longleaf forests stretched on for hundreds of miles, at first glance, an unvarying landscape dominated by a single tree species. But on closer inspection, these longleaf prairies are the richest forest type in North America, populated with more species of plants than any other ecosystem on the continent. Most of those plants seldom grow higher than your knees, but their variety is astounding.

On a spring morning at my favorite spot at Splinter

aged to turn the tables on the animal world, drawing most of their nutrients from insects they kill and digest. At Splinter Hill, tens of thousands of white-topped pitcher plants are visible at one time, their heads bobbing in the morning breeze, their pitchers full of the death that sustains the entire system.

The tall, fluted pitchers are actually the leaves of the plant, which grow into slender tubes capable of holding liquid. A combination of rainwater collected in the tube and enzymes and nectar secreted by the plant create a sort of witch's brew in the pitcher that draws various insects close. Lured by the smorgasbord of scents, insects climb, fall, or fly down into the pitcher, splashing into the liquid. The vertical walls of the pitcher, lined with downward pointing hairs and covered in a waxy substance, conspire to prevent escape, until the struggling insects—love bugs, moths, lady bugs, crickets, spiders—drown and are digested.

"Weird plants! Wonderful plants! Plants that eat insects and even animals! One of the great centers for carnivorous diversity in the world is sitting right here on the edges of the Delta," said Bill Finch, author of *Longleaf, Far as the Eye Can See* and science advisor to the Mobile Botanical Gardens.

"There are in Alabama nine species of pitcher plants alone. That is the world's greatest concentration of pitcher plants anywhere, and you know, we have a tremendous number of pitcher plant species that exist only along this Alabama corridor. We've got what I call the copper top pitcher plant, the cane break pitcher plant, the green pitcher plant, the parrot leaf, the frog's britches, and now we have a new species, Sarracenia rosea, that's been hiding in plain sight all this time. The list goes on and on, and there are a bunch more carnivorous plants."

Sundews and thread dews grow close to the ground; their leaves evolved into arms coated in glands the produce a substance sticky enough to ensnare even the largest dragonflies. The various bladderworts make their living using ingenious underground traps that snap shut when tickled by tiny creatures moving through the wet soil. One species of bladderwort actually floats in open water in the swamps of Splinter Hill and the Delta, able to trap larger prey such as mosquito larvae, baby fish, and tadpoles. Imagine—a plant that eats fish! The butterwort family catches its dinner by ensnaring gnats and other creatures on leathery leaves coated in something very similar to Vaseline.

It is all these meat eaters that drive the bog economy. Built on soils too poor to nourish most plant species, the bogs rely on the population of meat eaters to make up the difference. By harvesting the energy contained in the insect world, the carnivorous plants, when they die or burn in fires, complete a cycle that makes up for the impoverished soil. In turn, those soils are able to support an array of grasses, orchids and other plants unprecedented in the nation.

"I think Splinter Hill probably has somewhere in the range of twenty carnivorous plants on that one site," Finch said. "That's incredible diversity. It's hard to explain, except that these longleaf forests and this habitat have been around and fairly stable for much

A tree frog clings to an unopened orchid blossom, probably hunting the katydids that eat the orchids.

Above: an ocean of pitcher plants in a Delta bog. Far right: The pearl crescent is one of the "brushy-footed" butterflies. The brushy foots, which include the Gulf frittilary (right), are so named because their tiny front legs look almost like little brushes. They are common across the South.

longer than most habitats in North America."

In the late 1960s, something remarkable and now almost forgotten happened in one of the Delta's pitcher plant bogs in Baldwin County. A scientist with Auburn University, George Folkerts, decided to count all the flowering plants to be found in a square meter of ground. His findings were astounding. In that square meter, he counted sixty-three species, a stunning display of biodiversity for a piece of land in Alabama. Or, as it turns out, for anywhere in the world.

While a patch of grassland prairie in Romania was recently discovered to have eighty-nine species in a square meter, the Baldwin County total of sixty-three still ranks among the highest ever recorded on the planet. Sadly, the Baldwin County spot is not what it once was. In fact, it is now a cow pasture. Unfortunately, that has been the fate of the lion's share of Alabama's pitcher plant bogs. They've become pastures or tree farms or subdivisions, lost before we even knew what was in them.

A mess of love bugs, beetles, moths and ants inside the stem of a pitcher. All of these bugs fell for the alluring mix of nectar and the smell of death. Those at the bottom of the pitcher are further along in the decay process, already nourishing the plant, while some at the top are still alive.

Just as birds migrate across the ocean, so do fish, congregating along the rich waters of the Alabama coast in massive numbers from spring through fall, feasting on the nutrient-rich ecosystem fed by the rivers of America's Amazon. Cigar minnows are common baitfish. They are members of the jack family. *Dr. Bob Shipp's Guide to the Fishes of the Gulf of Mexico* notes that as bait, they bring $7 a pound, one of the highest prices of all Gulf fish.

7

The Fertile Crescent

All the water running off all the land into all the rivers in the Mobile River Basin is headed to one place: Mobile Bay. There, it mixes with the salty waters of the Gulf of Mexico and creates one of the most abundant coastal areas in the world, responsible for a sizeable portion of the seafood supply in the United States.

To understand the system, think of the Mobile River Basin and its forests as a giant biological machine that transforms abundant rain and sunlight into biomass, meaning animals and lots of plants—from the tallest trees to algae in the water. As those creatures and plants live and die, their wastes and carcasses decompose and return energy to the land. Every year, all that stored energy, collected everywhere from the floodplain forests to the smallest creeks, is swept downstream with the winter and spring floods. Fed by the nutrient-rich rivers of the Mobile Basin, the Mobile-Tensaw Delta is the engine that fires the coastal ecosystem of the northern Gulf of Mexico. The Delta absorbs all the energy flowing into it and, with the spring floods, gives it back at just the right moment.

The cycle begins offshore in the Gulf of Mexico. By late January and February, the marine world is feeling amorous. Snapper and grouper, amberjack and blue crabs, mullet and armies of shrimp gather in huge groups. For some species, such as mullet and snapper, these assignations occur in secluded spots thirty miles offshore. For others, such as sheepshead and blue crabs, the spawning takes place within site of land. Either way, the game is the same, find creatures that look like you and simultaneously cast your eggs and milt into the great blue sea.

Within hours of fertilization, the eggs of marine creatures become larvae, most capable of some form of locomotion. As they swim, the egg yolk sac is consumed. Within days, a baby sheepshead starts to look like an adult sheepshead, and a baby menhaden starts to look like a menhaden, and so on. By February, the Gulf is swarming with billions of larvae, representing the young of every creature that swims in the sea.

Favorable currents sweep the larvae, still so small they are measured in millimeters, toward shore. Coincidentally, all those tiny bodies—shrimp and fish smaller than a grain of

Mobile Bay as Nursery Habitat for the "Fertile Fisheries Crescent"

Clockwise from top left: Hermit crabs are the garbage men of the estuary, forever scavenging the bay's detritus. Gulf pipefish are perfectly suited for a life hunting and hiding in the grass beds. A tiny brittle star clings to a sea whip soft coral at Perdido Pass. A baby striped burrfish, better known as a porcupine fish, will grow up to have a body covered in needle-like spikes. Known as glass minnows, or sometimes rain minnows, massive schools of anchovies show up in Mobile Bay in early spring. A pair of mangrove snapper, one two months old, one two weeks old, and a school of tiny speckled trout, all caught while both hiding and hunting in a sea grass meadow. Tiny larval shrimp, just days old, wash into Alabama's estuaries beginning in February. By June, they have grown to market size. A school of sardines shelters under a raft of sargassum a few miles off the Alabama coast.

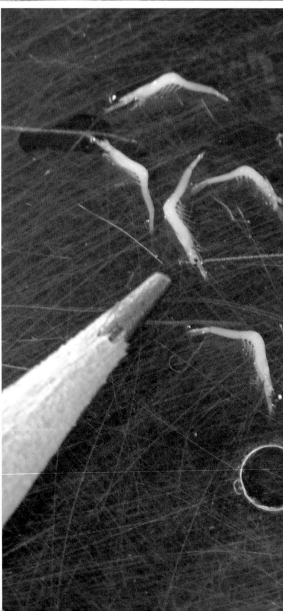

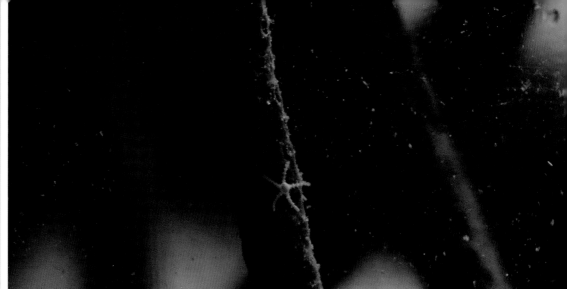

rice—arrive in the inshore nursery grounds just as the spring floods are pouring their riches into the Delta and the marshes of Mobile Bay and the Mississippi Sound. The influx of nutrients feeds the growth of the plankton and algae that fish and shrimp larvae will consume in their first summer. Those young creatures are, in turn, eaten by everything else, from birds to redfish to dolphins.

This is the hatching time for the marine world, a period of explosive growth that rivals any in the natural world. For the next six months, the waters off of Alabama will be among the

Round scad and Atlantic bumper migrate along the Alabama coast and into the mouth of Mobile Bay every spring and fall in massive schools that are sometimes more than a mile long.

richest on the globe thanks to the massive jolt of energy pumped out of the Mobile-Tensaw Delta. In fact, the portion of the Gulf Coast beginning at Mobile Bay has been named the Fertile Fisheries Crescent by biologists, richer even than the area off Louisiana plumped by the Mississippi's output.

"From Mobile Bay to the coast of Louisiana is probably the richest area in the northern Gulf, and the northern Gulf is one of the richest marine ecosystems in the world," said Bob Shipp, retired

head of marine sciences at the University of South Alabama. "The nutrients coming out of the Mobile Basin translate into fish. The fish here grow faster; there are more of them; there is more prey. It's just an incredible system."

But it is also a fractured system, one that man has altered in dramatic ways. Left to its own devices, Mother Nature would send a huge amount of energy from the Gulf back up the rivers in the form of migrating fish. Once upon a time, the Mobile Basin was home to massive migrations of almost two dozen species of fish that would head upstream to spawn, just like salmon on the Pacific coast. But locks and dams built in the last seventy years have broken that energy connection, which used to link the mountains to the sea with a two-way highway. Today, energy only flows downhill.

Making matters worse, energy is not the only thing flowing down the rivers of the Mobile Basin these days. By the time the rivers reach the Delta, they are carrying a heavy load of industrial and agricultural pollutants, including poorly treated sewage from hundreds of small towns. The state's long-standing approach to growing water pollution has been to rely on the old maxim "dilution is the solution to pollution" to protect its rivers. The inescapable signs that this approach is failing can be found all around Mobile Bay.

In 2009, fish all over the bay were covered in bloody lesions. Scientists with Auburn University linked the outbreak to pesticides carried in the spring floods, which were extra large that year. And every summer, dead zones build up in Mobile Bay, areas where the water doesn't have enough oxygen to support marine life. The dead zones are directly tied to the fertilizers flowing in from the rest of the state. Scientists have begun to talk about Mobile Bay approaching a tipping point,

A baby sargassum crab clings to a clump of floating seaweed, which makes up one of the main ecosystems in the Gulf of Mexico.

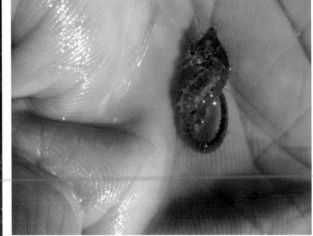

Above left: A pair of dwarf seahorses in an Alabama grass bed. They mate for life and grow to be about an inch and a half long. *Above right:* An adult male dwarf seahorse pictured in the author's palm. *Right:* A mohawk-sporting Molly Miller blenny peeks from an empty barnacle on an oil platform at the mouth of Mobile Bay.

just as they did around Chesapeake Bay in the decade before that system collapsed.

You can see the same signs of decline in the woods of the Delta or in the dwindling bits of native forest and stream left in the upper reaches of the state. The extinct animals are never coming back. And more creatures that call our hills and valleys and creeks home are surely headed for the history books.

For those that know these Alabama forests and rivers the best, every trip into the woods or a bog or the sprawling Delta is bittersweet. For all the glories to behold, in birds and fish and flowers, there is the nagging thought of what once was.

William Wordsworth, who never set foot in Alabama, published his poem "Intimations of Immortality" in 1807, but he might have been writing about this moment in Alabama's natural history. We would do well to pay attention:

> Though nothing can bring back the hour
> Of splendor in the grass, of glory in the flower;
> We will grieve not, rather find
> Strength in what remains behind.

Alabama stands at a defining moment. We still have the most incredible river system in North America. We are still home to more species than any other place on the continent. But the clock is ticking.

Silversides are common in Mobile Bay. They grow to about six inches. The splash of orange behind the eye identifies these as tidewater silversides, versus Atlantic silversides. They eat zooplankton, larval shrimp, and anything else they can fit in their mouths.

Left: A Native American shell midden on the Tensaw River, likely thousands of years old, and in near continual use until the 1700 or 1800s. In this part of the Delta, the middens are made of rangia clam shells, heaped up to make mounds several feet above the water. They provided a dry space to sleep and cook in a swampy world where most of the land was soggy mud. This particular site appears to have been a pottery factory. *Right:* Shards are everywhere mixed among the shells, likely things that broke during the firing process. Across the river is one of the tallest bluffs overlooking the Delta. It is composed of layers of rich marine clay that is still prized today by modern potters.

8

Harnessing Nature's Bounty

Human exploitation of the Mobile River Basin probably began with a mussel or a fish, some little morsel plucked from the water because it was good to eat.

From that moment, perhaps fifteen thousand years ago, the story of man in this place we call Alabama has centered around this river basin and what people could cull from its waters, forests, and mineral deposits. And so it goes in the modern age. Our use of these rivers has changed everything about them, including their appearance, their depth, the purity of the water running in them, the creatures that live in them, and when and where we let them flow. Dammed, channelized, fished, poisoned, and used as a superhighway for industry—we have had a one-sided affair with our rivers, all taking and no giving.

It is hard to conceptualize the bounty that would have met the first native peoples who happened upon an Alabama river in all its pre-industrial glory fifteen thousand years ago. Shoals of mussels, many containing a million or more individual mussels, would have been scattered in every reach of every river, from the top of the state to the bottom. Taken together, they would have acted as a water filter of almost unfathomable power, capable of filtering all the water moving through the state several times over every day. There are historical descriptions of river water so clear you could see to the bottom ten feet down. Compare that to the one-foot or less visibility typical in most Alabama rivers today for an inkling of just how much has changed.

The fish populations in those clear and undammed rivers would have been stunning. In a world before commercial fishing, and before our use of the rivers fractured the ancient spawning runs of more than a dozen species, fish migrated up and down the rivers with the seasons just as birds move around the country to escape winter cold. The first explorers wrote of rivers teeming with fish, and for anyone who has seen the redhorse suckers spawning in the Mobile Basin tributaries today, or the massive schools of threadfin shad or herring in the Delta, such tales are easy to believe.

THE OLDEST SIGNS of the rivers' first occupants are the shell middens. Scattered north and south, the middens represent early encampments from stone-age residents. They were places where thousands of mussel shells were heaped up to make a dry spot above the swampy bottomland surrounding the rivers. They are the last remnants of a people who lived barely above—and entirely off of—the rich waters of the Mobile Basin.

Cruising through the rivers of the Mobile-Tensaw Delta today, you catch glimpses of shell middens peeking out along the banks, a flash of starch-white shells rising above the black mud of the surrounding floodplain. Stop for a moment and search among the shells and you'll often find a bit of pottery, sometimes with geometric patterns cut into the ancient clay. The pottery is flecked with bits of mussel shell, which added strength to the bowls and cups, and is also proof of a fairly advanced culture. Most of the middens, which number in the hundreds, have never been explored by archaeologists.

Rich in calcium from the disintegrating shells, the middens are home to an unusual assortment of plants, including red buckeyes, which are seldom present in other locations on the floodplain. Archaeologists believe the highly toxic buckeyes were cultivated on the middens for use as a fish poison. Simply grind up the nuts, make little balls out of the paste, toss them

A cluster of wolfberries hangs heavy with fruit on an oyster shell midden in Grand Bay along the Mississippi line.

into slow moving water and wait for the fish that eat them to float to the surface.

The notion that Native Americans were gardening on man-made islands scattered throughout the Mobile River Basin is borne out in other plants present there today.

"The middens are relics. They are five-thousand-year-old gardens," said Bill Finch, a historian of natural systems and science advisor for the Mobile Botanical Gardens. "The stuff you find on middens you almost can't find anywhere else. It's very odd. But they are all plants that fruit, or there are all these weird little herbs you don't see other places around here but have medicinal uses. These things were planted in these places on purpose."

Finch has made a hobby of searching out remote shell middens in the Mobile-Tensaw Delta and documenting the strange plant assemblages that persist today.

"There's a weird tree yucca that you simply don't see except on these shell middens. It has a delicious fruit. There are several species of odd hackberries—the dwarf hackberry, the iguana hackberry—that don't exist anywhere, but they produce a delicious fruit that's great to eat," Finch said. "Then there's that weird, weird vine, Sageritia. I almost call it vining huckleberry. The fruits taste like huckleberry, but it flowers in the winter and fruits in April. They are very rare vines. I have never seen them except on a shell mound. All of these plants you see growing on the shell mounds because people brought them there. And in some cases, the middens are the last refuge for some of these rare plants."

Closer to the coast, the middens are made of oyster shells instead of mussels. There are dozens of these low, windswept islands that barely rise above the sea scattered around Grand Bay and Portersville Bay in Alabama's portion of the Mississippi Sound, and some along the shores of Mobile Bay itself. The oyster middens have their own strange plants, more tolerant of salt and repeated immersion in brackish waters. Perhaps the most spectacular plant is the wolfberry, a brilliant red tomato relative sometimes called the Christmas berry.

The plants fruit in December, hence the name, or perhaps it stems from the way the ruby globes stand out against the grays and greens of the wintertime marsh like ornaments on a Christmas tree. Unknown to most by dint of growing primarily on these remote and inhospitable islands, the berries pop in the mouth with a taste of salted tomato, a pinch of sugar, and a flash of spice. The texture of the flesh most closely resembles a small grape tomato.

It is possible to sit today on one of the ancient oyster shell middens eating a freshly shucked oyster topped with a ripe wolfberry and imagine that someone might have sat in the same spot five thousand years ago doing the very same thing.

FOR THE MOST PART the middens are about thirty to forty feet long, large enough to accommodate a dozen people as a primitive campsite. But in truth, even with their prehistoric gardens, the middens are small potatoes compared to the archaeological ruins found elsewhere in the Mobile River Basin.

To really understand how important these rivers were to the peoples living around them one thousand years ago, take a trip to Moundville, on the banks of the Black Warrior River near Tuscaloosa, or better yet, get in a canoe and visit Mound Island, one of the largest and least explored archaeological sites in the United States.

Both ancient cities were built by members of a tribe known to historians as the Mississippians, though in recent years it has become clear that the Mobile River Basin, not the Mississippi River, was the heart of the tribe's world. At each Alabama location, giant mounds rise from the earth, some more than fifty feet

The author descending from the summit of the largest mound at the Mound Island complex. (Courtesy of Bill Starling)

tall, some large enough to hold several football football fields. Moundville is the earlier site, believed to have been established about 1000 AD, and Mound Island was established around 1250 AD. It is believed both settlements provided homes to up to one thousand people, with perhaps ten thousand more living in nearby encampments surrounding each location.

Perched on a bend along Bottle Creek in the heart of the Mobile-Tensaw Delta, eighteen mounds rise up from the muddy floodplain on Mound Island, their sides so steep that walking up them is difficult. Historians believe the members of the tribe called the island "Mobila." The tallest mound stands eighty-one feet above sea level and provides a commanding view of the surrounding forest. It is believed that the chief who lived atop this mound controlled most of the Gulf coastal plain between modern-day Pensacola and New Orleans.

The Mound Island group, known today as Mobilians, held "a sort of primacy in religion, over all the other nations in this part of Florida," French historian Pierre-François-Xavier de Charlevoix wrote in 1722, when the Mound peoples were in the beginning of a decline that led to their end in the 1800s. Other historians have noted the importance of the site as well.

falcons and other animals. The tribe also held a complex view of the spiritual world, with visions of an afterlife and of animals as beings that were able to move back and forth between the worlds of the living and the dead. Waselkov curated an excellent exhibit on the Mound Island culture that can be seen at the University of South Alabama's Archaeology Museum. The oldest pieces in the collection date back about eight thousand years and include stone spear points and other weapons from the basin's earliest inhabitants.

Waselkov described the engineering employed by the Mound Builders as fairly advanced, with a lot of earth-moving accomplished solely with baskets woven from cane plants and hoes made of sticks and shells. Archaeological digs have revealed that the people who lived on the island ate a lot of animals, including a lot of shellfish, and a lot of corn. Buildings constructed on top of the mounds included posts set into the earth.

"This was a big city for the time. It was the center of the local culture," Waselkov said.

The Mound Builders had a good thing going in the Mobile River Basin, right up until the arrival of European explorers. Writing in the National Park Service's survey of the river system, Waselkov pinned the beginning of the decline of the native tribes and their cities to the first contact with the Spanish.

"Hernando de Soto's failed effort to conquer southeastern North America for Spain had broad and long-lasting repercussions for the region's indigenous peoples and their environment.

The streams flowing into the Delta are bursting with life. This section of one of the coastal creeks contained 26 species of fish in a half-mile stretch, more species than can be found in the entire Colorado River basin, which drains 11 states.

"When the sacred fire of any community in the vicinity went out, representatives made the sojourn to Mobila, where they rekindled it," wrote Robbie Ethridge, a University of Mississippi professor, in her book *From Chicaza to Chickasaw: The European Invasion and the Transformation of the Mississippian World, 1540-1715*. "People for miles around had considered the old mound center at Bottle Creek, which Iberville visited in 1702, as the 'place where their gods are, about which all neighboring nations make such a fuss and to which the Mobilians used to come and offer sacrifices.'"

Greg Waselkov, a University of South Alabama professor who has participated in excavations on Mound Island, said the tribe's art was exquisite and included human effigies and intricate carvings of peregrine

Even though the climactic battle at Mabila, where Chief Tascalusa ambushed and nearly destroyed the conquistadores in 1540, took place probably some distance north of our study area," Waselkov wrote, "Native societies peripheral to the invasion route experienced population declines, collapse of chiefly power structures, an end to earthen platform mound construction, and abandonment of many agricultural fields, towns, and hunting territories."

Indeed, by the time the French explorer Pierre Le Moyne d'Iberville's expedition visited Mound Island one hundred fifty years later, he described it as "an old village that is destroyed." But, the natives leading the French explorations of the Mobile-Tensaw Delta still considered Mound Island "the place where their gods are."

Some historical accounts suggest a remnant few kept the sacred fire alive on Mound Island into the 1800s. But the tribe faded away as the rivers became increasingly important to the growth of the burgeoning cities of the white man being constructed on their banks.

As Alabama was settled, several now-famous naturalists visited the forests of the Mobile River Basin, then considered remote and unexplored. Traveling along its rivers, they reveled in an array of plants and animals unlike anything yet seen in Europe or even in the cities of the Atlantic seaboard.

The first and most famous visitor was William Bartram, who explored the Mobile-Tensaw Delta in about 1776. The Delta's two-hundred-mile long Bartram Canoe Trail, the longest canoe and kayak trail in the nation, was named in honor of his explorations. Botanists around the country know Bartram best for his eloquent, poetic descriptions of the flora and fauna he encountered.

Describing the lowly members of the bream family, which includes bluegill, longear sunfish, and largemouth bass, Bartram wrote, "the top of the head and back, of an olive green, besprinkled with russet specks; the sides of a sea green, inclining to azure . . . The belly is of a bright scarlet red, or vermillion, darting up rays of fiery streaks into the pearl on each side."

Bartram described more than a hundred new plant species on his travels through Alabama and parts of Florida in the years surrounding the American Revolution. Many of those species ended up being cultivated in other states as ornamental garden plants, and hundreds of specimens he discovered and sent to England in his lifetime are still preserved in the British Museum's international collection.

"I was struck with surprise at the appearance of a blooming plant, gilded with the richest golden yellow . . . perhaps the most

A longear sunfish. (Courtesy of Bill Starling)

Right, Bartram's primrose has become a popular garden plant all over the world due to its enormous flowers, delicious scent, and copious blooming. *Below*, a rose gentian is a glistening jewel along a secluded bayou off the Alabama River.

pompous and brilliant herbaceous plant yet known to exist," Bartram wrote upon encountering one of the largest members of the primrose family—a Baldwin County, Alabama, plant that has been cultivated in the gardens of the northeast as "Bartram's primrose" since its discovery more than two hundred years ago.

For my money, the work of a more recent traveler makes for the better read. Phillip Henry Gosse, perhaps best known as the father of the modern aquarium, arrived in Alabama in the mid 1800s. Following in Bartram's footsteps, and inspired by his work, Gosse brought a painter's eye and a poet's heart to his 1859 book, *Letters from Alabama on Natural History*. By then, Mobile, Montgomery, and other cities had reached such size that they were beginning to have an effect on the local environment—most notably because much of their waste was simply thrown in whatever river was closest, as Gosse points out on several occasions.

Of his arrival at Mobile's fetid port after traveling up the bay from the Gulf, Gosse wrote:

> I was surprised to observe dead horses and cows suffered to lie exposed on the shore, scarce out of town, a neglect which I should suppose by no means likely in this hot climate to contribute to the health of the inhabitants.
>
> The exhalations arising from the extensive muddy flats, which are left uncovered at low water, must likewise be very prejudicial, and probably materially tend to give this town the unhealthy reputation which it possesses.

Mobile, "unhealthy reputation" and all, was a city on the move back then. Ambling along the old wharves, down where the Alabama State Docks are these days, in pre-Civil War Mobile would have seen just about every type of vessel afloat, from slave ships to cotton steamers to boats loaded with European immigrants. Just before Gosse arrived, the *Mobile Register* documented the arrival of the ship *Eliza Morrison*, captained by a John McCullock and loaded with seventy-nine Irish immigrants. Among them were farmers, laborers, children, and women, including one Jane Stain—listed in the ship logs as a "spinster" though she was but fifteen years old. The chatter of immigrants, the calls of oystermen hawking their catch, the hiss of steam boilers firing up amidst the clanking of anchor chains and the screeching of seagulls were the sounds of Mobile's wharves.

Mobile was second only to New Orleans for the tonnage of cotton that moved through the port during those years, with more than thirteen million bales passing through during the heyday of King Cotton, from 1817 to 1861. It is a sure bet that almost every wisp traveled downstream from the Black Belt plantations via steamboats cruising down the rivers. In those days, before seventeen dams turned the rivers into a series of big lakes bordered by vacation homes, things looked a lot different. Mighty rivers like the Alabama or the Black Warrior were broad and shallow, with riffles, sandbars, and rocky shoals throughout. Almost as soon as he left Mobile and passed north into the river system, Gosse's writing took on a different flavor. Behold his romantic description of one of the Mobile Basin's damselflies:

> He who would see the Emerald Virgin must go to some such hidden brook as I have described; over which as it flows silently—in a deep soft bed of moss of the richest green, or brawls over a

pebbly bottom with impotent rage—three or four of these lovely insects may be seen at any hour . . . a fly of surpassing elegance and beauty, whose long and slender body is of a metallic green so refulgent that no color can convey any idea of it.

While historians more often note Gosse's discomfort after witnessing slavery at its most brutish during his travels, his descriptions of the forests he passed as he journeyed upriver in May of 1838 are unrivaled:

> The banks are clothed with tall forests to the water's edge: trees arrayed in all shades of green, of various height and form, some covered with glorious flowers, suddenly appeared and as swiftly vanished, a constantly shifting panorama . . . Beautiful flowers, of varied colors and fragrant perfume thronged the edges

The Emerald Virgin.

The forest floor on the Delta's edge is often dominated by a carpet of palmetto fronds. It is hard to imagine a more primordial spot.

of the forest. . . . The opening of the spring is the most interesting season of the year, when after a suspension, more or less absolute, of activity and life, all nature springs into fresh existence: the gate of Eden is, as it were, reopened, and birds, insects, and flowers, renew their Creator's praise.

Gosse summed up his spring journey upriver with a regret that he had arrived in May, when the season was already well underway, bordering on summer: "However, be it mine to notice what still remains to be observed, instead of regretting that which is past and cannot be recalled."

Such is our lot when studying the Alabama forests that surround us, for the forests of today are not the forests Henry Gosse saw before the Civil War. Those magnificent longleaf and cypress forests—the forests of our great, great-grandparents—those forests have been cut down. And the forests that grew back in their place, those have been cut down, too.

Old Alabama was a land of giant trees. You can see that in the ancient cypress forest found fifteen miles offshore in the Gulf of Mexico. A remnant of the Mobile-Tensaw Delta from an age when sea level was three hundred or more feet lower than it is today, some of the cypress stumps in this long-submerged forest measure ten feet across. Those trees provide a time traveler's peek at what the state's forests looked like before people began cutting them down one hundred fifty years ago. Both the longleaf and the cypress trees were so big that it was possible to stand a team of mules on top of one of the stumps and turn them in a circle, according to the exquisite book *Longleaf, As Far As the Eye Can See.*

One such tree remains, hidden away on an island bordered by a bayou named Jessamine, just below Devil's Bend. It is the lord of the Delta, a cypress tree that measures twenty-seven feet around and is somewhere between three hundred and five hundred years old. This particular tree is hollow, and is missing its top, perhaps due to a lightning strike. That damage is most likely the only thing that saved it from the woodsman's ax. Being hollow, the tree wouldn't have had the highly valuable heartwood so prized for shipbuilding. It simply wasn't worth the effort to cut it down, so the loggers left it standing.

Robert Lesley Smith, born in 1918, remembered the big trees, the logging of the swamps, and a rough life along the river firsthand. In his time, logs were cut in the Delta after high water, once the ground had firmed up enough to support ox teams. The huge trees were then floated out when the floods returned the following spring.

"My father was a logging man," said Smith, author of *Gone to the Swamp*. "I loved to follow my father about. He made his entire living from the swamp and logging. He owned ox teams and employed a couple of dozen workers at a time. . . . it was hard, back-breaking work. Men were killed by falling trees, and so much of the work was done on the water, many good men drowned in the process of floating logs out in high water. I remember a number of good friends who drowned here and there along the river."

The so-called "watermen" were the highest paid members of the logging crew. They danced upon the logs as they were hauled through the swamp and kept them spinning with their feet and long poles, which aided the passage of the giant trees through the muck and mire.

"There was so much brush in the way, the logs had to be kept spinning at all times. The waterman did that. By the force of his body, he kept that log moving. It was a highly prized skill, and they were paid twice as much as anybody else. But it was dangerous," Smith said. "A waterman was as respected in the community as an automobile mechanic, and mechanics were almost worshipped in those days. You would speak of someone and say, 'He's a waterman, you know.'"

As the years went by and the trees closest to the edges of the rivers were cut, loggers had to work on higher ground, farther from the water. Smith watched as

the technology used to retrieve the timber evolved from a pair of oxen to an eight-wheeled wagon hauled by four pairs of oxen to pull boats, which used winches to pull trees through the swamp, carving long trenches that dramatically altered the shape of islands and the forests on them.

"Then the era of the truck came along, and when the season was right you could get trucks into the Delta and do some of the things the ox were doing. Eventually, things progressed to where they were using helicopters in the 1960s and '70s."

In the end, just about every inch of the Mobile-Tensaw Delta was logged. Study an aerial photo of

Lord of the Delta, the last of the ancient giants—this cypress, estimated to be as much as 500 years old, is 27 feet around. It can be reached only with a muddy hike through knee-deep swamp water.

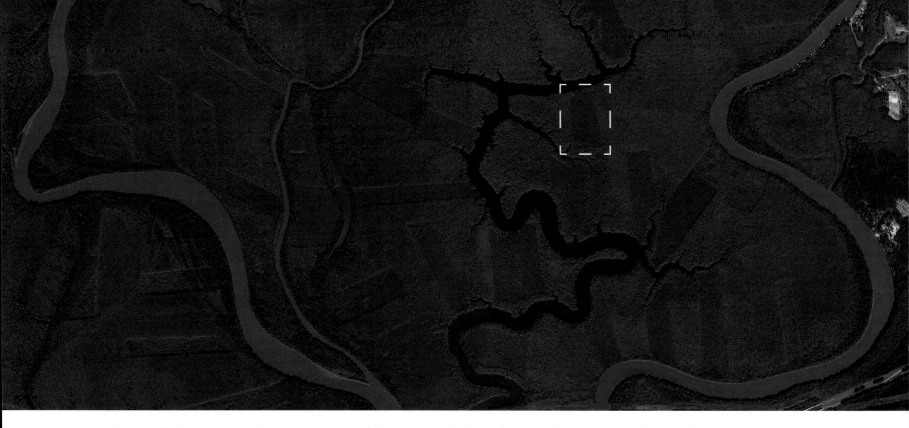

This satellite image depicts clear-cut logging scars (with an example indicated) surrounding and on top of Mound Island, home to one of the most important and least explored Native American sites in the Southeast. These cuts, made in the 1980s–90s, were harvested using helicopters to fly the old-growth cypress logs to lumber mills. (Google Earth, Data SIO, NOAA, U.S. Navy, NGA, GEBCO Image Landsat / Copernicus Image © 2020 TerraMetrics)

the region today, and you will see so many healing clear cuts that it looks like someone made passes across the land with a giant lawn mower.

Shortly before he died at age ninety-nine in 2017, Smith said of the forest he remembered from his youth: "Ninety-nine percent of it is gone. It will regenerate, I believe. I certainly won't see the day, but it will regenerate. . . . With what I saw in my early days, that's just a wonderful thing. I just hope I live to see it taken care of. I hope we don't make the same mistake and tear it up again."

The logging Smith witnessed—which still goes on today, with Alabama ranking third in production of timber nationally, after Oregon and Georgia—was part of the first in a wave of significant alterations to the Mobile River Basin's landscape in the last one hundred fifty years. It coincided with the rise of King Coal in Alabama's natural history.

Legend has it that coal was discovered in Alabama in the early 1800s by boys on a hunting expedition. They pitched camp and fell asleep around their fire, only to awaken when the stones ringing their campfire burst into flames. Though the tale may be apocryphal, it is still possible to walk through the woods surrounding the headwaters of the Mobile Basin and find loose chunks of coal scattered like rocks. The area surrounding the Black Warrior, the Cahaba, and the Sipsey rivers is famous in the geological and metallurgical world as the only place on Earth you can find all the ingredients to make pig iron—coal, iron ore, and limestone—in such close proximity.

The first coal industry in the state involved farmers and blacksmiths who dug their own coal for heating homes or firing forges, but it quickly became a fully mechanized industrial excavation that rivaled any Victorian age industry in terms of human and environmental cost. The legacy of mining Alabama

coal, and of stripping bare the forests that overlay the richest coal seams, is a sad one, full of death and destruction—for the miners and in the natural world surrounding every mine.

As early as 1886, assistant state geologist Henry McCalley estimated that the Warrior Coal Field alone, around the Black Warrior River, contained "a sum total of coal of not less than 113,119,000,000 tons"—that's 113 billion tons of coal in just one of numerous coal fields in north Alabama.

McCalley also made a case for the first of the great alterations man would make on Alabama's rivers in order to use the Mobile Basin waterways as an industrial pack mule:

> This field and this county have, winding through their most productive areas for miles, a river that can be made navigable for steam tugs and coal barges all the year round, with a minimum channel, at extreme low water, of eighty feet wide by four feet deep, for the sum of $400,000 to $1,200,000, according to the nature of the work. Congress has already made a large appropriation for the opening of this river, and will doubtless continue in this good work until a waterway is opened to the sea for the cheap coal of the Warrior Field.

At the time, the north Alabama coalfields were yielding more than four thousand tons of coal a day, a small fraction of what they would one day produce. Mobile, a few hundred miles downriver, provided a port to export the haul to the world, and McCalley became the great champion of altering the rivers to facilitate barge traffic. He was paid by the federal War Department to "ascertain the nature and extent of the obstructions to navigation and to obtain estimates of the cost of removing or overcoming the same." McCalley was also tasked with cataloging the natural resources along the riverbanks. He identified one of the state's most valuable commodities as

On the edge of the Delta sits the enormous coal ash pond at Alabama Power's Barry Steam Plant. Spring flooding in 2020 came within a few feet of overtopping its levee. A breach here like the one in Kingston, Tennessee, would flood across the entire Delta, killing everything in its path. Incidentally, the billion gallons of ash recovered from the 2008 Tennessee failure ended up in an Alabama landfill, because Alabama was the only state willing to accept the toxic waste. (Courtesy of Cade Kistler, Mobile Baykeeper)

the longleaf and cypress forests, "especially in the hollows and ravines along the water courses."

"The Warrior and Cahaba Coal Fields also have along their streams some of the grandest and most picturesque of scenery, and innumerable sites of narrow channels, with rocky bottoms and sides, for the erection of machinery of any magnitude," McCalley wrote, at once acknowledging both the beauty of the area and the fact that these lovely river valleys were destined for other purposes. Indeed, in those same pages, he laments a world lost amid the burgeoning mines.

"The county between Jones Valley and the Warrior River, and on the south side to the Cahaba River, was full of game . . . The Warrior and Cahaba rivers were beautiful streams, clear as crystal, in which you could see a fish in ten feet of water," McCalley wrote, describing a vanished world we can scarcely imagine today. "The fisherman in his canoe, dug out of a poplar tree, with his gig in his hand and his rifle lying beside him, ready for a deer if he should venture in sight, with the muscadine vines hanging in festoons from the tops of the tall trees that overhung the water with their clusters of black, delicious fruit, and the beautiful red-horse fish sporting beneath his canoe, with their silver sides and red fins and tails, in the most desirable and healthful climate in the United States."

McCalley cautioned that, "What I have said of Jefferson County in the foregoing pages is intended to apply from its earliest settlement up to the year 1861." The transformation McCalley would have witnessed between 1861 and his death in 1904 for these "grandest and most picturesque" river valleys would have been profound, especially as modern environmental protection laws were still a century away. With the high sulfur and arsenic content in the coal coming from Alabama mines came heavy environmental damage, including reports of trees dead for miles around mining operations, due to the acidic air pollution coming from the coke ovens and steam engines powering the underground work.

The water was polluted on a similar scale, due partly to the exceptionally high arsenic levels in Alabama coal, but also to the wastes generated by thousands of men working in remote locations for years at a time. As early as 1924, an Alabama farmer filed suit against a coal mine near his farm for polluting a stream on his property with "human excrement, coal washing, coal ashes, cinders, and poisoned foul matters." Other lawsuits from the era describe streams so polluted that no fish survived in them, so poisonous that any livestock that watered in them died. There are no reliable estimates of how many creatures may have slipped into extinction during the early years of coal mining in Alabama.

Controversy over coal mining in Alabama, particularly the environmental costs versus the economic benefit, still rages today. While various politicians campaigned for reelection in 2014 using the slogan "Coal Jobs Count," the city council of Birmingham was part of a coalition fighting the creation of a 1,700-acre strip mine on University of Alabama land on a bend in the Black Warrior River. The city council's primary objection was that one of the outfalls for mining waste flowing into the river would be adjacent to the drinking water source for hundreds of thousands of people.

The vision of rivers as industrial conveyance described by McCalley in the late nineteenth century has come to fruition thanks to numerous dams and locks on Alabama's rivers. Even today, coal remains one of the main cargoes moving up and down the rivers. Coal imported by Alabama Power from Colombia makes its way up the rivers to the company's coal-fired power plants, while the coal mined in Alabama is shipped downriver for use overseas. Most Alabama coal is sent abroad today, due in part to the high-sulfur content, which means it pollutes more than the low-sulfur coal—from Wyoming, Colombia, and other places—that is burned by the state's utilities today. Such shipping would be impossible without the numerous dams on every one of Alabama's major rivers.

Sadly, the staggering ecological toll of those dams didn't become apparent for decades, and now it may be too late to reverse the damage.

Right: Miller's Ferry Lock and Dam was completed in 1969. (Google Earth, Data SIO, NOAA, U.S. Navy, NGA, GEBCO Image Landsat / Copernicus Image © 2020 TerraMetrics) *Below:* Two years later, the final nail in the coffin of the ancient fish migrations, Claiborne Lock and Dam, the last of 30 major dams built on Alabama's rivers, reduced the most diverse river basin in North America to a single free-flowing section about sixty miles long. (Courtesy U.S. Army Corps of Engineers)

9

Dam It All!

Millions of years in the making, the great fish migrations of the Mobile Basin ended the year man first walked on the moon. The annual spawning runs of more than a dozen Mobile Basin species were sliced in half when the gates of the Miller's Ferry Lock and Dam clamped shut on the Alabama River in 1969. The problem was compounded with the addition of the Claiborne Lock and Dam farther downstream two years later. The annual migrations had suffered a crippling blow years earlier, in 1960, when the Tombigbee was dammed, but so long as the Alabama and its tributaries were still connected to the Gulf, fish could move upstream and spawn.

By 1971, the great rivers of the state—the Tombigbee, the Alabama, the Coosa, the Tallapoosa, the Cahaba, the Black Warrior, which together played host to the greatest collection of aquatic diversity in the hemisphere—had been reduced to a single free-flowing section about sixty miles long that encompassed the system's Mobile-Tensaw Delta and not much else. Fish that had migrated up and down the state's rivers for eons could no longer make the trip.

The Alabama and the Tombigbee were the Mobile Basin's main arteries, with thousands of smaller rivers, creeks, and tiny brooks branching off like capillaries in the human bloodstream. After countless insults to the rivers over the preceding two hundred years, those dams finally severed the ancient connection between the tiniest tributary streams in the top of the watershed and the Gulf of Mexico. That connection, linking the mountains to the sea, was really about energy. All the biological energy generated by the plants and animals in the watershed was gathered every year and delivered to points downstream with the annual floods. But that flow was suddenly cut off by the dams, starving the entire system of nutrients, floodplain-building sediments, and—for many creatures—the basic ability to reproduce.

But that energy exchange also flowed uphill, in the form of the millions of fish that migrated upstream, their oily flesh carrying massive stores of energy gathered out in the open ocean, or in the brackish lower reaches of the estuary and the energy-rich Delta. Just as salmon on the West Coast feed and

A single throw with the castnet during the fall menhaden migration in the Mobile River. These are young of the year fish. Born in February, they pass from egg to larvae to juveniles by April, and then grow explosively through the summer, converting the energy coming down the rivers into flesh.

fertilize the upper reaches of their home rivers with energy from the ocean when they die after spawning, so did the fish that migrated up Alabama's rivers.

Mooneyes, striped mullet, striped bass, menhaden, the American eel, needlefish, Alabama Shad, skipjack herring, Gulf sturgeon, paddlefish, and a dozen other species all made spectacular runs up and down the rivers before the dams.

For instance, both mullet and the Alabama shad migrated all the way from the Gulf to the Cahaba River around Birmingham or to the Black Warrior around Tuscaloosa, more than three hundred miles inland. Those migrations, which involved millions of fish, ended almost overnight when the dams came in. A few mullet still make it up the Cahaba, but they are measured in individual fish these days, not in the millions of pounds the way they used to be. And the Alabama shad, which once supported a thriving commercial fishery in the rivers, has become so rare it was recently labeled a "Species of Greatest Conservation Need" by federal officials. The best hope for survival of Alabama's namesake shad now lies in Florida's Apalachicola River, where they are able to migrate about a hundred miles upriver before hitting a dam.

SOME OF THE MIGRATING fish, like the sturgeon, were gigantic, seven feet long and weighing hundreds of pounds. Prior to the dams, the sturgeon would head upriver every spring after fattening up all winter on shrimp, marine worms, and fish in the Gulf. The sturgeon would travel hundreds of miles inland to the streams where they were born to spawn, a journey repeated in these rivers since dinosaurs roamed the hills and valleys of Alabama. There are still people alive who recall seeing these veritable monsters wiggling their way up shallow riffles to their spawning grounds.

"A friend of mine and me were fishing with poles down on Cahaba River. Out in the middle of the river, this thing raised up with all those ridges on its back. We threw our poles down and run and told everybody but nobody would believe us," said Jannie Falkenberry, relating the catching of the largest fish recorded in an Alabama river, a 341-pound sturgeon captured in 1941. "But later, we saw the fish again, the big sturgeon. It got up between the shoals and the river went down and it was caught between the shoals. So that's the way we found it."

Leslie Buzbee's family has owned Buzbee Fish Camp on Bay Minette Creek for nearly a century. He lived on the property his entire life, until passing away in 2016 at ninety-one. I interviewed Buzbee in 2012. He spoke of watching hundreds of sturgeon at a time passing up the rivers and creeks, right up until the dams were put in.

"In spawning time, you'd see just hundreds and hundreds of sturgeon, just butting that lock, butting that lock, trying to get further up, further up, until their whole head would be bloody. You knew what was happening. They can't get where they want to go to lay," Buzbee said. "There ain't no more sturgeon here no more. And there ain't gonna be."

Above: The fish Jannie Falkenberry saw trapped between riffles in the Cahaba River. At 348 pounds and 7 feet long, it was the largest freshwater fish ever caught in Alabama (*Centreville Press*). *Left*: Children pose with a Gulf sturgeon caught in the Coosa River in the 1940s. By the end of the 1960s, nine dams blocked access to the section of the river where these fish spawned. (Elmore County Historical Society)

Bottom: This massive turbine brought electricity to many when Martin Dam went into service in 1926 (Alabama Power Co. Archives). It also blocked access to the Tallapoosa River for millions of anadromous fish that spawned there. *Top:* This twin waterfall, fifty feet high, was lost with the closing of the dam on Smith Lake, one example of the disappearance of thousands of creeks and rivers behind 30 major dams in the state.

Alabama even had its own species of sturgeon, the Alabama sturgeon—at three feet long, a pint-sized version of its mammoth cousin—that made annual migrations up the rivers. While scientists believed the last Alabama sturgeon died in 2012, a victim of the dams blocking the spawning runs, a new DNA sampling technique suggests the species may be hanging on in a few places and could possibly be saved. Fish, like all animals, including humans, constantly shed their DNA, in everything from scales to eggs to bodily waste. That DNA can persist in the water column at the cellular level for several weeks, depending on exposure to sun, high temperatures, and other factors.

Much like a crime scene investigator swabbing surfaces for blood, hair, or fingerprints, scientists are now able to use state-of-the-art genetic techniques to search for organisms living in lakes and rivers.

Alexis Janosik, an assistant professor at the University of West Florida, and Steve Rider, a biologist with the Alabama Department of Conservation and Natural Resources, collected 130 water samples up and down the Alabama River in 2014 and 2015. A handful of those samples contained traces of DNA believed to belong to the Alabama sturgeon. It is the most hopeful sign yet for a species of fish thought to be lost. The concern now is that the DNA might come from the last members of the race, solitary fish isolated in the big lakes above the dams and unable to make their spawning migrations, or even find the last of their kin.

"Their native spawning runs have been fractured by our use of the rivers," said Pat O'Neil with the Geological Survey of Alabama. "Many times, these fish come back to their natal streams to spawn. It's not only the sturgeon that do that. The shad, the skipjack, the Alabama shad—they would mount these large runs up the rivers, but they have to have fresh water in order to spawn. That's their life cycle. It's one of the more interesting aspects of coastal biology. But it is also why these fish are rather imperiled now."

WHILE THE TWO DAMS on the lower Alabama River were the final nails in the coffin for Alabama's great fish migrations, construction on the proverbial coffin began much earlier, in 1914, to be precise.

That's when a fledging Alabama Power Company got its start with Lay Dam, the first hydroelectric dam in the state, and the first dam on the Coosa River. By the 1960s, seven dams were constructed on the Coosa, and today only a few miles of the river remain in something resembling a natural state. The old river has been turned into a series of stagnant, murky reservoirs. The plight of the river was featured in a 2009 report, titled "Seven Sins of Dam Building," by the World Wildlife Fund, which documented what the group called some of the most damaging dams on the planet.

"The Coosa River was once one of the most biologically diverse rivers in the world, but today it is the most developed river in Alabama," reads the World Wildlife Fund report. "The damming of the Coosa has been described by the U.S. Fish and Wildlife Service as [causing] 'one of the largest extinction rates in North America during the 20th century.'"

Indeed, biologists link the damming of the Coosa River directly to the extinction of forty species, describing it as the single greatest cause of U.S. extinctions. Many of those lost species were mussels and snails, creatures that thrive in the shallow and rocky riffles flooded by the dams. Even today, the river is still quite diverse, home to 54 species of mussels, 110 species of snails, and 147 species of fish. But some of those creatures are on the verge of oblivion, already listed as endangered species. It is a sure bet some of them will ultimately be lost to extinction.

The U.S. Army Corps of Engineers released a new environmental impact statement for the fourteen major dams on the Coosa, Tallapoosa, and Alabama rivers in 2014, as part of the federal relicensing of the dams' permits, which occurs every thirty years. Three of the dams are operated by the Corps and eleven are operated by Alabama Power Company. Describing the effect of the dams, the report acknowledges the "decline or loss of river-dependent species of freshwater fishes, mussels, and snails," and it states that, "habitat fragmentation effects of dams in [this portion of the Mobile Basin] have resulted in declines in habitat for anadromous fishes." Anadromous fishes migrate between fresh and salt water as part of their life cycle.

The damages on the Coosa are profound and ongoing. One section of the Coosa is known to environmentalists and locals as the "Dead River" because Alabama Power has been allowed to let oxygen levels in the water routinely drop below the threshold needed to support life. U.S. Fish and Wildlife Service scientists, who surveyed the river as part of the relicensing process, say that several threatened or endangered species native to the river, such as painted rocksnails, tulotuma snails, and southern clubshell mussels, have been totally wiped out in that section.

In the end, the impact statement called for no significant changes to the way the dams operate, despite the known effects of the dams on creatures protected by the Endangered Species Act.

After a thirteen-year process, the federal government reauthorized the Coosa dams in spite of clear evidence that they were continuing to cause the extinction of more and more species. Compare this to major changes required on the West Coast—such as the addition of fish ladders to help salmon make it upstream to spawn—when dams there were relicensed. You can be sure the federal government has never required a fish ladder on any of the thirty dams in Alabama.

It was truly stunning, then, when a federal appeals court ruled in 2018 in favor of the Southern Environmental Law Center (SELC) in a lawsuit that sought to have Alabama Power's

Right: The lure of the pocketbook mussel is on display in the Coosa River as the mussel tries to trick a fish into carrying its larva upstream. (Courtesy Paul Freeman, The Nature Conservancy Alabama chapter) *Below:* The Southern combshell mussel uses a small, subtle lure to attract a log perch. When the fish tries to eat the lure, the mussel clamps shut on the fish's head and shoots larvae into the gills, for transport upstream. (Courtesy of Paul Johnson, Alabama Aquatic Biodiversity Center)

new thirty-year operating license for the dams revoked on the grounds that the license didn't do enough to protect the Coosa river and its inhabitants. Ultimately, the suit was successful, said Gil Rogers, an SELC lawyer, because the new license—like the old one it was to replace—violated multiple federal laws. The group argued that the license lacked protections for endangered aquatic creatures and habitat, environmental studies required by law, and a requirement that Alabama Power ensure that oxygen levels in the river remained high enough to sustain life.

It will take several years to rewrite a new, more environmentally sensitive operating license. In the meantime, Alabama Power and federal officials have already agreed to restore minimal flows to the "Dead River" section of the Coosa to better support fish and mussel life. Even so, the river will never be what it was before the dams.

"Dams have brought many good things to Alabama, including improved river navigation, economic development, and hydropower, but their impact on biodiversity has been and continues to be devastating," said Scot Duncan, head of the biology department at Birmingham-Southern College and author of *Southern Wonder*, an excellent book about Alabama's biodiversity.

"Dams transform shallow and oxygen-rich river ecosystems into deep stagnant lakes, and block the natural movements of animals, sediments, and nutrients. Very few river species can survive in the reservoirs created by dams. Dams are why most of our extinct and endangered species in the state are aquatic animals," Duncan said.

You can see the biological loss caused by the dams most clearly in the Mobile Basin's mussel population, which E. O. Wilson describes as "one of the richest and most imperiled assemblages in North America." Some Alabama species had already gone extinct by the 1800s, according to historical accounts, due in part to the discovery of pearls in some freshwater species. Another wave of destruction hit the state's mussels between 1900 and 1940, when Alabama mussels were used as a raw material to make buttons; the state's mussels were especially prized due to the wide variety of colors present in the shells. The rise of plastic buttons after the 1940s put an end to the button harvest.

But the biggest culprit in the ongoing decline of the state's mussel fauna, are the dams, said Paul Johnson, head of the Alabama Aquatic Biodiversity Center, which specializes in rearing rare mussel and snail species in its laboratory in Marion for reintroduction around the state. Johnson said it is almost impossible to calculate how many mussels were present in the state prior to the dams. Every river and creek was home to huge shoals of mussels. Early travelers wrote of floating over multicolored river bottoms carpeted with mussels that glistened in shades of green, blue, purple, orange, and brown.

The dams delivered a one-two punch: They flooded the shallow, pebble-bottomed riffles where most

A handful of hatchery-reared orangenacre ladyfisher mussels are individualy labeled so scientists can study their survival after they are planted in a riffle in Tallatchee Creek.

93

Jeff Powell, with the U.S. Fish and Wildlife Service, and Andy and Paul Johnson with the Alabama Aquatic Biodiversity Center, plant orangenacre mussels just above a riffle in Tallatchee Creek.

mussels prefer to live, and they cut off the fish migrations, which have recently been revealed to be critical to the survival of the mussels.

At first blush, a mussel appears to be little more than a lump of meat sitting in an armored box, passively filtering water and siphoning out a meal. There would seem to be little connection between mussels sitting on the bottom of a stream and fish swimming over them. The reality, however, is quite different and provides a fascinating example of evolutionary adaptation.

For mussels, sedentary animals living in a river where the current is always flowing downstream, reproduction presents an unusual challenge. If mussels were to spawn like oysters—or like most fish species, for that matter—with males and females simply casting their milt and eggs into the current, the babies would always end up somewhere downstream. In just a few generations, the mussels, sliding ever southward, would run out of river and be crammed up against the inhospitable Gulf of Mexico.

Mussels clearly had to come up with a different strategy to

ensure that some of their progeny made it upstream. So they became fishermen.

The mussels' life cycle depends entirely on their ability to lure passing fish to take their various baits. Study for a moment the endangered orangenacre mucket, a rare mussel with a caramel-colored shell. Its scientific name, *Hamiota perovalis*, translates loosely as "lady fisher." At spawning time, the lady fisher, like many of the state's species, produces a superconglutinate, which is a larvae packet. This is where it gets interesting.

The female lady fisher extends this larvae packet out into the current on a nearly invisible strand of mucus, which can measure up to eight feet long. The larvae packet dangling on the end of this invisible fishing line looks for all the world like a small minnow. It is cream-colored, about two inches long, with a black stripe down the side, just like a fish's lateral line. Completing the costume, the larval packet even has a dark spot on each side that looks like a minnow's eye.

Quivering in the current, the fake minnow is impossible for a passing fish, such as a bass, to resist. Once a fish has taken the

94

bait, the larval packet pops open in the fish's mouth. But instead of a meal, the fish finds itself with a bunch of baby mussels adhered to its gills. Those larval mussels will ride around for several weeks before dropping to the river bottom and settling in to a new home. Timed with the annual fish migrations, the mussels evolved a surefire method of making sure some of their offspring landed upstream.

Each of the state's 182 species of mussels evolved different lures and slightly different strategies. Some angle for a specific species of migrating fish. Some larval packets resemble crayfish. Other packets move seductively, undulating like a fish digging for food on the river bottom. Members of the kidneyshell family cast their larval packets out into the current to settle onto rocks. There, the packets are taken up by fish, which mistake them for the fly and insect larvae they perfectly resemble.

All of this fancy trickery worked well for millions of years, resulting in the world's greatest diversity of mussel species. In fact, more than 60 percent of all mussel species in North America live in Alabama, and the Mobile Basin has been acknowledged as the global center for mussel diversity since the 1800s.

Then came the dams. For the mussels, whose entire reproductive strategy is predicated on the annual upstream migrations of millions upon millions of fish, the arrival of each of the dozens of dams—spread across every one of Alabama's major rivers—was simply catastrophic.

The state's mussels bear wonderfully whimsical names, including the monkeyface, snuffbox, fuzzy pigtoe, heelsplitter, wartyback, pearlshell, fatmucket, hickorynut, elktoe, spectaclecase, orangefoot pimpleback, iridescent lilliput, finelined pocketbook, fawnsfoot, and turgid blossom. But in every case, mussels face an uncertain future due to the presence of the dams. The lady fisher, or orangenacre mucket, for

instance, was listed as endangered by federal officials as early as 1993, just twenty-four years after the arrival of the dams on the lower river.

"Of the 182 mussel species that live in Alabama, there are twelve that can live in lakes," said Paul Johnson, describing the transition that takes place when a dam is installed—from free-flowing stream to slow-moving lake—as fatal for most mussel species.

When their home waters are lost among the drowned streams and valleys behind a dam, the mussels simply die, Johnson said. Then, any mussel remaining in the headwaters of a stream that runs into one of the dammed lakes finds itself biologically isolated, as the fish that might spread its young to other places cannot traverse the dams and the big impoundments. Now, scientists fear a new wave of extinctions among these isolated relic mussel populations as the mussels—which can live over a hundred years—are slowly dying off from old age.

Today Johnson acts as a veritable Johnny Appleseed when it comes to preserving the state's mussels. The Aquatic Biodiversity Center, which Johnson has headed since its creation in 2005, has a single mission: To rear and reintroduce the state's rare mussels, snails, and fish to the places they used to live. Carrying a handful of individually numbered orangenacre muckets from his truck to a small stream for planting, he called them nature's water filter.

"There were millions of these mussels in every river all over the state, each one capable of filtering a gallon or two an hour. These animals create your clean water," Johnson said. "So, when you had densities in literally the thousands or millions all over the state, they literally were the most important component of our ecosystem."

Johnson called preservation of the state's mussels critical to the future health of the rivers. With mussel reintroductions underway since 2005, perhaps the state's mussel diversity will be preserved for another generation, and maybe, just maybe, that will be long enough for something dramatic to take place on Alabama's rivers—the removal of some of the dams or a return of some of the migrating fish.

At this point, the dams have been in place so long they have created their own economy, with each of the dozens of giant lakes created by the dams serving as major recreation destinations. Those lakes are surrounded by thousands of vacation homes, fishing marinas, and boat dealers. Assuming that the majority of the dams will remain in place, going forward the best option is probably to try and lessen the impact of the dams on the river systems.

From the perspective of the creatures in the rivers, we compound the negative impact of the dams on the Coosa and the Tallapoosa, the Sipsey and the Cahaba, and the rest of the rivers by the way we use them. In effect, the dams function like a savings account for water, storing the high flows during the rainy winter and spring so that water can be used during the hottest and driest part of the year for hydropower production and to keep the river channels deep enough for barge traffic.

But that's the opposite of the way plant and animal life adapted in this incredibly rainy spot. The Mobile Basin's flora and fauna evolved in a system with predictable seasons, including a season of high water in winter and a season of low water in the summer.

Mitch Reid, with the Alabama chapter of the Nature Conservancy, explained:

> Impoundments operated for hydropower fundamentally alter the flow of the rivers such that the system essentially runs as the inverse of its natural flow. Natural periods of high water, typically during our winter and spring months, are held back and then released during the summer, which is when the river should be experiencing its lowest flows.

This is done in order to maximize hydropower during peak demands, when Alabama is cranking the AC, but it wreaks havoc on the system. Those natural high flows are critical to the spawning cycle of everything from the Alabama bass in the Alabama River to the oysters in Mobile Bay, and those low flows are needed to allow for the river to reset and resist invasive or noxious weeds and pests.

Those alterations in flow are most visible in the Mobile-Tensaw Delta, where the interaction between salty Gulf water and the flow from the rivers is most critical. Creatures such as white shrimp, menhaden, mullet, and a hundred other species have timed their spawning to be in tune with the arrival of the spring floods. Likewise, other creatures have come to depend on the push of salt water up into the Delta and its rivers in late summer. Oysters, for instance, need regular fluctuations between salt and fresh water in order to battle internal parasites and predatory snails.

In the impact statement on the dams, the Army Corps of Engineers acknowledged that "flow manipulation undoubtedly adds to the cumulative effect of existing anthropogenic stressors" on Mobile Bay and the rest of the system, and it concludes that "the best strategy for protecting the ecology and biodiversity of the basin, including its protected species, is to maintain or restore to some extent the natural patterns of variability of flow regimes throughout the basin."

But the natural flow patterns would have the rivers at their lowest precisely when the dams are releasing the most water, to meet demands for navigation and hydropower in late summer. Compounding the issue is the influence of the Tombigbee River, which—since its man-made connection to the Tennessee River and the creation of the Tenn-Tom Waterway for expanded barge traffic—has become something of a wild card in managing freshwater flows in the Delta, apparently adding more to the system than expected.

In the battle over what nature needs and what man demands from Alabama's rivers, nature has always been the loser.

WHILE IT IS HARD to envision the U.S. Army Corps of Engineers removing the dams at Miller's Ferry and Claiborne—which ensure navigable channels on the rivers—some progress is already being seen with the removal of smaller dams on the Alabama River's tributaries.

Case in point is Marvel Slab, a six-foot dam on the Cahaba River, constructed around 1960 as a river crossing for trucks hauling coal and logs. Though it was only six feet high, it cut off all upstream migration in the upper Cahaba. The low dam, like hundreds more scattered around the state, was a relic of industries that had long since moved on from the area, but it was still taking a biological toll. Working together, the Corps, the Alabama Department of Conservation and Natural Resources, the Nature Conservancy, and the Cahaba River Society removed the dam in 2004. It was the first significant dam removal in the state's history. Today, seven endangered species, including two fish, two mussels, and three snails, are holding their own in the area surrounding the old dam.

Paul Freeman, an aquatic ecologist with the Nature Conservancy, was part of the group that removed Marvel Slab. Today, that same coalition has set its sights on a much bigger target,

namely, restoring the lost migrations past Miller's Ferry and the Claiborne Lock and Dam.

The general idea was to mimic the function of the salmon ladders, built at a cost of millions of dollars per dam, on the rivers of the Pacific Northwest. Those fish ladders create a small raceway of stair-step pools that allow migrating salmon to swim past the dam. Lacking the money, the pressure from the public, or the federal will to install similar devices on the Alabama dams, Freeman and Steve Herrington, also with the Nature Conservancy, came up with a solution.

Instead of a fish ladder, why not use the existing

The Nature Conservancy's Paul Freeman holds a paddlefish that has rubbed its snout raw against the concrete dam trying to find a way around it to get upstream for spawning.

locks already in place at the dams? The locks move barges and other river traffic around the dam by closing the gates at each end of the lock and raising or lowering the water inside to meet the level of the river above or below the dam. They could be used the same way to move fish. The biologists found that the sound of flowing water, such as a trickle from a hose splashing into the lock area, was enough to coax fish into the lock.

"In 2009, the U.S. Army Corps of Engineers began to open the navigation locks for fish passage during specific times of high migration. This 'conservation locking' is intended to move

Top: This paddlefish, recovered during electroshocking, lost its entire snout. It was most likely broken off after the fish got caught in a hydraulic current at the base of the dam. *Right:* A team from the Nature Conservancy, U.S. Fish & Wildlife, and the Alabama Department of Conservation and Natural Resources, surgically implants a transmitter in the belly of a paddlefish to follow its efforts to migrate upstream.

fish past the dams so they can reproduce, as well as provide ecological benefits for the Gulf and the contributing rivers," Freeman said.

I witnessed a team of scientists conduct an electroshocking session inside a closed lock at the Claiborne Dam the year the conservation locking experiment began. Electroshocking relies on a boat rigged with a generator and an array of probes that emit a non-lethal shock underwater. When the fish get hit with the voltage, they are temporarily stunned and float to the surface, where scientists can net them, tag them, or study them. After a few moments, the fish wake up unharmed. Claiborne Dam is located about eighty miles inland, a purely freshwater environment. Yet there in the lock, among the species present were a number of saltwater fish, including flounder, croaker, and striped mullet, a testament to the fact that Gulf fish once moved far upriver in the Mobile Basin.

In addition, there were a number of three- and four-foot long paddlefish in the lock. Paddlefish—which are sometimes called spoonbill catfish by Alabama anglers—are bizarre and ancient, older even than the dinosaurs, according to the fossil record. They swim with their bucket-sized mouths agape, sieving the water for tiny organisms. The name stems from the long, paddle-shaped snouts they sport, which are used to sense plankton in the water.

I will never forget the first paddlefish I ever laid eyes on. Its snout was rubbed raw and bloody from trying to push its way past the gigantic concrete dam. Another had its snout broken clean off. That fish, Freeman said, got caught in a violent current at the base of the dam. Its snout was snapped off when the fish was slammed into the river bottom headfirst.

Several universities, state and federal agencies, and the Nature Conservancy have been monitoring fish passage through the locks. So far, the conservation locking appears to be successful, though whether it can restore meaningful numbers of migratory fish remains to be seen. "From 2010-2012, 171 fish representing eleven migratory fish species were captured downstream of the dams, tagged with transmitters, and then released. The tags contained sonic beacons that were used to track the movement of the fish," Freeman said. "Nine of the eleven species entered the lock chamber during this study and several species were tracked long distances, up to 157 miles, as they swam upstream."

Conservation locking is one of the new elements recommended as part of the Army Corps of Engineers' environmental impact statement, which calls for operating the locks twice a day, once at 8 a.m. and once at 4 p.m. I would argue that we could do much better and open each lock at least four times a day, perhaps six. The additional cost for the extra openings would be a fraction of the money spent to restore salmon migrations in the Pacific Northwest.

But some fish passage is better than no fish passage, and the creation of the Alabama Aquatic Biodiversity Center and the fledging conservation locking program are bright spots in the sad tale of our use of Alabama's rivers. But it is hard to imagine that opening the locks a few times a day can ever make up for the unimpeded passage of a dozen species, and all the energy they carried, up and down the rivers.

It is even harder to imagine how our lost rivers would have looked each year in springtime, as millions of silvery bodies shimmered over the endless rainbow shoals of mussels.

The road next to the high school in McIntosh, Alabama, was paved with mercury waste, seen in this image. For decades, Olin Corporation gave its deadly wastes to residents for use as driveways and to communities for roads and in parks. After fifty years of polluting the community and nearby river, the waste was finally cleaned up.

10

A Delta Dimmed

The rivers of the Mobile-Tensaw Delta have cast their spell on generation after generation. For most who fall for the purling song of the rivers, it is a lifelong love affair, full of endless spring mornings and long summer afternoons. Like all lovers, these self-described river rats always remember best their first encounters with this watery world.

Find yourself alone on a hidden slough as the morning mist rises and a flock of ibis wade toward you, plucking tadpoles from the shallows, and see if you can resist the spell of the Delta. For those new to these waters, they seem as wholesome, lovely, and rich as any on Earth. What could possibly be amiss in a place where a person could venture out and catch fifty largemouth bass in the morning and fifty speckled trout in the afternoon? By all accounts, the system remains incredibly productive, with its estuary yielding thousands of pounds of crabs, fish, and shrimp annually. When compared to the vanished abundance of other fractured and impoverished systems around the nation, such as Chesapeake Bay or any of the California river systems, the Mobile Basin's estuary seems to be in prime condition.

It is a testament to the glory of the Mobile-Tensaw Delta and the bay that their power of seduction is as strong today as it was one hundred fifty years ago. But talk to an old timer, anyone who has loved this Delta and Mobile Bay for decades, and you hear the same tale, again and again. The refrain is a sad one—you should have seen it then; it's not what it once was; we've got to save it.

The problems in the Mobile Basin as a whole all come to roost in the Mobile-Tensaw Delta and its estuary, Mobile Bay, which serves as a gathering point for all that comes downriver, good or ill. For those who have spent the most time on the water, the decline is obvious.

The challenge for the current generation is to find a way to break the cycle of decline and preserve what remains. The best place to start is by listening to those who witnessed the sins of the past.

"WE WERE SWAMP RATS. The boys, not the girls. I learned that when I went to school. I could take offense, or I could embrace it," said Robert Lesley Smith of his childhood on the edge of the Mobile-Tensaw Delta. "To me, the swamp wasn't a hellhole like some people saw it. The swamp was adventure. The swamp was beautiful. . . . I could identify just about every tree in the swamp by its leaf, and sometimes by the smell of its sap."

Smith remembered the rivers almost unsullied, before industry and a burgeoning population had their way. Fishing was as easy as kicking over a rotten log and putting something wiggly on a hook. Everything the family needed came from the floodplain forests.

When someone would find a tree with a big beehive

Not all the birds make it. Did this shell fail because of DDT? A small bit of golden yolk was enough to attract a stream of ants marching from a nearby anthill.

in it, several families would get together, and the men would fell the tree. There would be so much honey we would use washtubs to scoop it out. There'd be honey running out on the ground and into the river. It would make a slick on the water a mile downstream.

Water for drinking, you'd just drink out of the stream. I can remember many times my father would take his old felt hat and dip it in the stream and I would drink from his hat. There just wasn't as much pollution upstream in those days. That's why we got by with it. You wouldn't dream of doing that now.

Smith, who lived through an era that saw a hurtling destruction of Alabama's most sensitive places, said of his generation: "We literally stripped the Delta; we ruined it." Yet when I talked to him about the future, I found optimism, a feeling that maybe the tide of decline is slowing and perhaps could even be reversed.

From the late 1940s through the 1990s, pollution from chemical plants, military weapon factories, steel mills, paper mills, and overwhelmed municipal sewer systems escalated at a torrid pace. Into the 1990s, Mobile County, which includes about half of the Mobile-Tensaw Delta within its borders, ranked among the top ten counties nationally for chemical pollution emitted from factories. By the 1960s, Birmingham had fifteen steel mills going full bore, with air and water pollution on an epic scale. At the time, before the advent of modern environmental regulations, much of the pollution was legal.

You can see the legacy of all this industry in the plants and animals surrounding old industrial facilities along the Mobile Basin's rivers, many now on the federal Superfund list, a roll call of the nation's most polluted spots. There is a recurring theme among these Alabama Superfund sites when compared to Superfund sites in other states. Again and again, the cleanup

goals settled on by the polluters, the U.S. Environmental Protection Agency, and the state of Alabama are less protective here than in other states and often allow stunning levels of pollution to persist for decades.

Take the case of two Tombigbee River factories placed on the Superfund list in 1984. At the time, hundreds of acres of Delta floodplain swamp adjacent to the factories were identified as ranking among the most contaminated places in the United States for DDT and mercury. A bass caught there by state officials had the highest level of mercury in its flesh ever recorded in a living creature. Meanwhile, both DDT and mercury contamination were widespread downriver. The last time federal officials conducted Gulf-wide testing for DDT and mercury contamination in oysters, Mobile Bay oysters topped the list, with the contaminated swamp identified by federal officials as the primary source. Pelicans nesting in Mobile

Bay at the time still contained DDT just below the threshold for egg failure, though the chemical had not been manufactured in Alabama since the late 1960s.

Yet between 1984 and 2008, the only cleanup the EPA or the state required in those swamps was the removal of DDT-contaminated sediments for a fifteen-acre patch out of about three hundred acres that were similarly poisoned. Meanwhile, those swamps flooded for months at a time during high water every year, spreading DDT and mercury far and wide for decades. Finally, in the last two years, federal officials have announced that the two companies that did much of the polluting—Swiss-based Ciba-Geigy and Olin Corporation, based in Missouri—will be required to cover the contamination with a layer of clay and sand, a cheap and easy fix that a review of federal records suggests

Many pelican nests have three babies, or three unhatched eggs. You can tell one of these chicks is older than his siblings. Fuzz is just beginning to emerge. The chicks are extremely vulnerable to sun exposure at this stage.

A trio of juvenile white ibis totter on a log in Coon Neck Bend. They'll turn snow white as adults, and make their living picking through the shallows for worms, frogs and minnows.

would have been required in most states as soon as the pollution was discovered, thirty years ago. Indeed, U.S. Fish and Wildlife Service officials proposed that very fix again and again over the decades, only to be rebuffed by the state.

"I'VE BEEN COMING TO Chuckfee since about 1946, as soon as we could get gas after the war and as soon as Homeland Security of the time cleared the harbor and we could run boats up here," said Russell Ladd, an insurance executive whose family has owned a camp in the heart of the Delta on Chuckfee Bay for seventy years. The camp is accessible only by water. In the early years, Ladd's father owned a wooden thirty-eight-foot houseboat—the *African Queen*—that served as living quarters.

Today, Ladd has a multi-bedroom cabin built on stilts with electricity provided by generators. His earliest memories of the Delta involve food.

"Everything was cooked in fat. We would always have grits, and we'd have pouldeau [coot] stew. Pancakes as big as the skillet. Fried fish. You could drop a pole with a worm and catch hundreds of bream in an afternoon," Ladd said. "We'd catch crabs and boil them in a pot on the back of the boat and eat them while we watched football. It was wonderful."

He said the first sign of trouble was in the 1960s, when the ducks stopped coming. "They'd sit out in Chuckfee and feed all night, mainly widgeons and gray ducks, a few mallards, a few pintails, a minor few canvasbacks, some black ducks and plenty of pouldeau. That bay was just black with ducks. We'd sit out on the boat and watch them fly in, watch them feed. You'd hear them, 'prrp, prrp, prrp,' all night long. That's the noise they'd make. Then in the sixties, they were gone."

Ducks are still seen in the Delta today, but nothing like what Ladd recalls. Others confirm his remembrances.

"I've found my father's duck [hunting] diary going back to 1937. It says that at certain times, the whole area around Polecat Bay used to get dark with ducks overhead in the sky. You don't see that anymore," said Robert Meaher, whose family donated the Delta land where Meaher State Park is located. "The last canvasback duck I saw in Mobile Bay was in 1966."

There are likely several factors behind the decline of the ducks, but a study by the U.S. Fish and Wildlife Service of ducks killed in Alabama during the 1965 and 1966 hunting seasons described them as "notably high" in DDT, infamous for causing eggshell failure. Another federal study near a DDT factory in Alabama noted a 97 percent drop in the aquatic bird population between 1948 and 1959.

Loss of habitat is clearly another factor behind the decline in ducks. The area where Meaher's father wrote of ducks blacking out the sky, North Polecat Bay, was destroyed by the U.S. Army Corps of Engineers. The agency used the once vibrant bay as a dumping ground for sediment in order to make a turning basin for ships on the nearby Mobile River. Lower Polecat Bay, meanwhile, was destroyed by an Alcoa aluminum factory, which was allowed to use it as a dumping ground for bauxite wastes, including a giant spill of the hazardous material. Then, when Alcoa closed the factory in 1982, the company used a loophole in the state contract to avoid a promised cleanup of the six hundred-acre area, according to media accounts from the time. Those lost bays are now listed as "industrial ponds" on most maps.

"My father liked to have had a fit when they filled in North Polecat Bay," Meaher said, recalling the beautiful area. "No ducks there anymore."

Ladd listed Polecat as one of a string of grass-filled bays the ducks normally moved through, spending time in one before migrating a short distance to the next.

"Polecat Bay was deep and crystal clear," said Vince Dooley, a Mobile native and famed University of Georgia football coach. Dooley's family has owned an island in the Delta along the edge of Polecat Bay since before the Civil War. As a child, the coach visited his grandparents on the island, where members of his extended family lived without benefit of plumbing or electricity into the 1970s. The only way to reach the island was by boat.

"They would pick us up in the launch and take us across Polecat Bay. I remember how clear it was. You could see fish and grasses way down deep," Dooley said of his childhood in the early 1940s. On a recent trip, the coach found the old family homestead, which was destroyed by Hurricane Frederic in 1979. The wreck of the large boat the family routinely used to cross Polecat Bay was there, but the bay behind it is now so shallow you could walk across most of it without getting wet above the knee.

"BY THE 1990S, ALL of the grass seemed to disappear," Ladd said. "Before that, you could not run an outboard across Chuckfee after the middle of May without pulling your engine up and cleaning grass off the wheel. The bream aren't here like the used to be, or the bass. We've lost the grass, which was the structure they liked."

Indeed, scientists at the Dauphin Island Sea Lab believe that Alabama has lost more than 50 percent of its submerged grasses since the 1950s. A recent study by the Mobile Bay National Estuary Program suggests the state lost an additional

Abandoned homeplace of
Vince Dooley's grandparents
and the boat they used to
ferry goods and people back
and forth from their home on
an island in the Delta.

1,300 acres of seagrasses—or 22 percent of all that remained—between 2002 and 2009. Only about 5,248 acres of aquatic grasses remain today, according to that study, and nearly 70 percent of that lies in the Mobile-Tensaw Delta.

Complicating the story of the estuary's grass beds was the creation of the U.S. 98 Causeway, a seven-mile dirt dike built across the top half of Mobile Bay in 1926 in lieu of a bridge. The Causeway was the only way to cross Mobile Bay until 1978. Scientists are still trying to understand all the ways the roadway changed the system once it divided the huge swamp above it from the bay below it, choking the flow of tides and the rivers for almost a century. The roadway is implicated in the spread of invasive aquatic plants, such as milfoil, that have spread over much of the Delta, replacing the native grasses that fish depend on.

"The Causeway acts as a breakwater. The natural wave action that could have controlled these species from taking over no longer exists. That's why you only see tape grass and other [native] plants with big roots to the south of the Causeway. They can survive the waves," said John Valentine, director of the Dauphin Island Sea Lab, who advocates opening up the Causeway to allow Mobile Bay waves to wash into the Delta's lower reaches unimpeded. "In the really hot summer months, those dense canopies of milfoil drain all of the oxygen out of the water column every night."

Valentine said scientists are now studying whether fish leave the small bays above the Causeway due to the milfoil-induced oxygen crashes, perhaps contributing to the disappearance of the bass and bream Ladd described.

"There was grass all the way down the Eastern Shore. Huge grass beds," said Jimbo Meador, seventy-nine, who has lived on the water at Mobile Bay's Point Clear all his life. Meador began hunting and fishing in the Delta as a child. As a teen, he trapped fur-bearing animals in the swamps for extra income, and he sold what he caught from Mobile Bay to local seafood shops. As an adult, he ran shrimp boats and worked as a fishing guide. He spent his life on these waters.

"People don't know how clear Mobile Bay used to be. Those grasses trapped the sediment. Now, when the wind blows, it gets all stirred up, and it stays that way for days," Meador said. "That's what happened to the grass. The mud blocked the light, and the grass died."

The once-clear waters of Mobile Bay figured prominently in Meador's courtship of his wife: "Lynn and I were dating, and she would come out with me when I went spearfishing for sheepshead," Meador said. "She would be up in the boat with a rake scraping barnacles off the pilings. The sheepshead come around to eat those barnacles. And I'd be in the water with my spear. A girl who'll scrape barnacles for you, that was the real catch!"

For people familiar with Mobile Bay as it is today, the notion that anyone could spearfish in its now murky depths is almost unbelievable.

THE DISAPPEARANCE OF THE seagrasses in Mobile Bay is no mystery. The culprit was mud. Lots and lots of mud.

Plain old mud is probably the most destructive thing we dump in our rivers. The problem extends from Shades Mill Creek in Birmingham to tiny Joe's Branch in Spanish Fort and everywhere in between. Every time a forest is cleared for a subdivision or a tilled field gets rained on, every time we pave a new parking lot where woods used to stand, we are making it a little harder for the Mobile Basin to survive.

When the Skylab space station passed over Alabama in the 1970s and took pictures, one detail stood out—a giant plume of mud, which spread across Mobile Bay and down into the Gulf

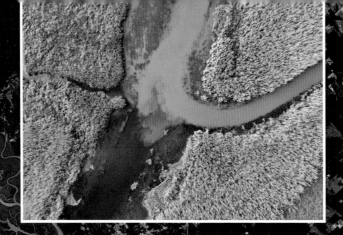

This photo, taken from Skylab space station in 1976, shows massive plumes of mud flowing from the Lake Forest subdivision on Mobile Bay. Construction of the giant subdivision took a decade and caused the complete destruction of seagrass beds on the east side of Mobile Bay. *Inset, top:* Similar muddy effect from a recent clear cut on the edge of the Delta flows unimpeded into the system, smothering everything in its path.

of Mexico. That plume emanated from Daphne, Alabama, and was caused by the creation of a sprawling subdivision called Lake Forest. So many millions of cubic yards of mud escaped from the massive construction site that it filled in D'Olive Bay in about three years, a process that would have taken thousands of years under natural conditions, according to geologists. The creation of the subdivision is widely regarded as the killing blow for the seagrass meadows of the Eastern Shore.

Remember that Alabama is one of the three rainiest states in the nation. It's why we have so many rivers. But it also means our rivers must deal with a lot of runoff. More and more, that rain is falling onto pavement. Instead of trickling through forests and soaking into the ground, the water is shunted straight to the closest creek or river, where it hits with the force of a fire hose or flash flood. Stream scientists call it headcutting. The increased water flows erode the stream banks, sinking the creek channel deeper into the landscape with every rain. You see it all over the state near malls and other developments. Streams that ought to be four or five inches below the forest floor are now in canyons eight or nine feet deep, carved by the sudden gush of runoff from storm sewers. As those streams sink deeper into the earth, they lose the wetlands that ought to be growing on their banks. And huge amounts of eroded mud and clay flow downstream.

All the mud flowing into the system chokes our rivers. It suffocates bottom-dwelling creatures like mussels and oysters in a layer of fine silt. The loss of mussel-provided natural filtering after the arrival of the dams upstate has only compounded the problem. Consider this: In as little as two weeks without light in mud-clouded water, submerged vegetation begins to die. In Alabama, our waters have been clouded with mud for forty years.

"What we have is a sediment problem. That's what wiped out the grasses," said John Dindo, a senior scientist with the Dauphin Island Sea Lab. "People don't realize that what's coming off that parking lot in Birmingham, or that construction site, that all ends up in Mobile Bay. What worked for our grandparents, just letting everything go downstream, that isn't working anymore."

Daphne, Alabama, bills itself as the "Jubilee City" on signs welcoming visitors to town. The jubilee is a phenomenon unique to the Eastern Shore of Mobile Bay, resulting in thousands of sea creatures crowding together in the early morning hours, right up against the bay shoreline on hot summer nights. It makes for an unforgettable tableau, the shallows carpeted in flounder and stingrays, with blue crabs and shrimp walking on top of them, and eels slithering amongst the group. Completing the portrait are dozens of people wading at four o'clock in the morning with flounder gigs and nets, scooping up the sudden bounty by lantern light.

There are historical records of jubilees from the 1800s, and some religious groups consider them a miracle from God, a gift from the sea delivered to the faithful. Scientists today understand their cause—pockets of low-oxygen water that get pushed ashore by currents and wind, driving sea creatures to the shallows in search of water with enough oxygen to support life. There is growing concern as jubilees become more common that often their cause is something more sinister. Raising alarm is that in the last four years, jubilees have suddenly become a regular occurrence on Mobile Bay's *western* shore, a historically unprecedented occurrence.

Jimbo Meador can predict jubilees with uncanny accuracy after a lifetime of studying the mix of recent rain, east winds, and slack tides that most often predicate an event. In the old days, he said, fish would come ashore in the night, but when the morning wind began stirring the water with waves, they would move back into the depths, none the worse for wear. Now, he says, fish often die in the jubilees.

"I see these mullet here, I've never seen mullet in a jubilee. I've never seen fish die in a jubilee growing up on the bay," Meador said during an early morning jubilee that was remarkable for the presence of tens of thousands of mullet schooled in the shallows and lots of dead fish along the shoreline. Mullet, living in the top of the water column, are not typically affected by low oxygen moving along the seafloor the way crabs, shrimp, and flounder are.

"Normally, everything lives after a jubilee," Meador said. "Now, I'm seeing dead eels on the beach. I'm seeing mullet up here I've never seen before. I think it's something else. Something was depleting the oxygen before, but this is something different for sure. . . . I think we are seeing algae blooms or something."

Valentine, with the Dauphin Island Sea Lab, has similar concerns.

"I don't get alarmed by the jubilees. Those are on record hundreds of years ago. The harmful algal blooms, those are things that tend to tell me something is wrong. We now have a lot of closed oyster grounds. Was it like that three hundred years ago? No. It is all the result of more infrastructure, more runoff, more sewage, lack of proper stewardship."

For the first time in more than a century, state officials permitted no harvest of oysters in 2018 because the reefs were simply too bare.

How far we've fallen. According to my calculations based on state harvest records, it is likely we have removed about 1.8 billion adult oysters from the bay in the last century alone.

Oysters were one of the primary attractions for the earliest Alabamians. Ancient shell middens in Bon Secour, Grand Bay,

on Dauphin Island, along Mobile Bay, and elsewhere attest to the importance of oysters as sustenance and as an architectural building tool. The native peoples actually lived atop the mounded shells of oysters they had eaten. The mounds, up to twenty feet high, helped the early coastal inhabitants get above sea level and the wet, soft ground along the coast. Testing shows that some of those mounds were heaped up more than six thousand years ago, and a decrease in the size of the shells deposited on those mounds most recently suggests that these earliest people were actually having a measurable effect on the reefs of Mobile Bay long before Europeans arrived here.

We have a more precise knowledge of the modern oyster harvest beginning in 1880, which was the first year the state kept records. That year, about 327,000 pounds of oysters were tonged from the bay. That is pounds of meat, not including the shell. The harvest quickly ramped up. Just a decade later, in 1890, the harvest reached 1.5 million pounds, and by 1928 it was up to 1.8 million pounds annually. To put that in perspective, a pound of oyster meat usually contains about fifteen oysters, according to estimates from federal officials, meaning the 1928 harvest consisted of around twenty-seven million oysters.

Between 1880 and today, records show that we have harvested at least eighty-five million pounds of oyster meat from our reefs (again, that weight doesn't include the heavy shells). Given that the state harvest data is missing for 50 out of 137 of those years, the true total is probably close to 120 million pounds harvested from our reefs, or about 1.8 billion live oysters. To really understand the problem facing us today, take a look at the average annual harvest totals beginning in the 1950s, compared to the harvest today.

Mobile's Shrinking Oyster Harvest

1950s: 1.3 million pounds

1970s: 756,495 pounds

1980s: 628,497 pounds

1990s: 594,137 pounds

2000s: 669,394 pounds

2010s: 142,359 pounds

2015: 33,586 pounds

2018: 0 pounds

Top: Shucking oysters a century ago in the Alabama Canning Company at Bayou La Batre, Alabama. The small boy on the left end is Mike Murphy, 10, from Baltimore. Bottom: Little Nettie, 8, made 45 cents a day as a shucker in 1911. The sharp-edged oyster shells have torn up her too-large shoes.

The harvest at the dawn of the twentieth century fed six canneries operating around the bay, including Graham Seafood Company, Hidenheim Company, McPhillips Packing Company, Marco Skremitti, Coffee Island, and the Alabama Oyster Company. In a sad bit of history, those factories operated largely thanks to plentiful child labor. In fact, Lewis Hine, a federal photographer tasked with documenting child labor practices for the government, captured numerous images of children as young as seven shucking oysters in Alabama canneries in February 1911. The kids worked for as little as thirty cents a day. In one devastating picture captured by Hine's camera, nearly the entire work force of the Alabama Oyster Company appears to be made up of school-age children. A few goonish adults are depicted, apparently a management team right out of the pages of Dickens.

This heyday of harvest for Mobile Bay oysters didn't last long. By 1950, there were just three canneries left, and by 1967, just one. "The decline of the canneries from their peak years of the 1920s was largely due to loss of productive oyster bottoms in Portersville Bay," reads the state history of the reefs.

That's a delicate way of saying we harvested all of the oysters from Portersville Bay in a few short years. Beginning in 1909, state officials began attempting to rebuild that reef by depositing baby oysters dredged from the nearby Mississippi Sound. In a harbinger of things to come, that attempt failed because the muddy bottom, minus its ancient covering of oyster shells, was too soft to support the oysters as they grew. They sank into the mud and suffocated.

That didn't stop state officials from continuing the practice.

By 1934, 117,000 barrels of oysters had been collected from the north and east sides of the bay and dumped on the reefs closest to the canneries surrounding Portersville Bay. Those oysters were collected via the use of oyster dredges, which are heavy metal boxes dragged along the seafloor behind a boat. The dredge plows through the reef, scooping up live oysters as it scrapes through and destroys the reef community they were growing on. A national study in 2012 pinpointed the use of dredges as causing "the functional extinction" of oyster reefs on the east and west coasts. In the words of Sean Powers, a senior scientist at the Dauphin Island Sea Lab, "where they've allowed dredging on wild reefs, the lesson is pretty clear. You will lose your oyster reef."

We've proven the truth of that lesson again and again on Mobile Bay.

But the worst damage done to the bay began in 1946, when the state opened the bay to the mining of old oyster and clamshell deposits for use as filler material in concrete. These shell deposits, dozens of feet thick, formed a hard rock-like matrix beneath the bay. Understanding why such thick layers were present requires a trip back in time to the last ice age, eighteen thousand years ago. Then, sea levels were more than three hundred feet lower than today, with most of the water on the planet frozen in huge glaciers that covered much of the land. Back then, Mobile Bay would have been a valley with a river flowing through its middle. As the seas began to rise at the end of the ice age, water poured back into Mobile Bay, and oysters and other creatures began to grow. Over millions of years, with a cycle of ice age followed by warm period occurring about every hundred thousand years, layer upon layer of oyster shells were deposited on the bottom, with new generations growing on top of the old shells.

The importance of these layers of shell to the integrity of the

seafloor in the bay cannot be overstated. Think of them as the foundation under a house. They provided a stable platform for new generations of shells to rest upon and ensured that those new shells didn't sink below the mud under their own weight as they grew. As those layers of shell were removed from the bay, more and more of the bottom was rendered too soft to support the weight of oysters.

An estimated forty million cubic yards of oyster shells were mined from Mobile Bay between 1946 and 1968. That's about four million dump-truck loads. A total of around eight million dump-truck loads of oyster shell were culled from the bay before the shell mining stopped in 1982. The state of Alabama reaped around $180,000 a year in royalties for this most destructive of practices. Worst of all, the practice was allowed to continue in Alabama for decades after Florida and Louisiana banned it in their waters in the 1960s.

A plume of thick mud fifteen hundred feet long oozed across the bottom away from the shell mining, covering and killing "everything in its path," according to a federal environmental assessment from the 1970s titled "Dredging of Dead Reef Shells, Mobile Bay." At the same time, a plume of mud twelve miles long was visible on the surface of the bay spreading away from the mining operation. The report described the conflict between the bay's oystermen and the mining company, acknowledging that many smaller live reefs were destroyed by the practice. It also concluded that "it is doubtful that oysters can be grown on or in the immediate vicinity of old dredge cuts within several generations." Nearly the entire floor of the bay was mined in this fashion.

The lingering effects of this mining can be seen every time the wind blows. Instead of a bay bottom underlain with a matrix of old shells that trapped sediment and supported live oysters, our bay is now home to a soupy layer of soft mud that oozes

between the toes of swimmers. Even a slight breeze stirs this layer into the water column. Anytime the wind blows stronger than about fifteen miles an hour, the water turns the color of chocolate milk, and stays that way for days.

The result is a murky bay where you can seldom see the bottom in more than about a foot of water. That's a far cry from stories told by people who remember what the water looked like even in the 1960s, when it was not uncommon to see the bottom in water as deep as eight feet.

The loss of thousands of acres of oyster habitat also means the loss of thousands of acres of critical nursery habitat for all aquatic creatures, for at some point in their various life cycles, almost everything that swims finds shelter in an oyster reef.

DESPITE WHAT HAS BEEN lost, and despite all we've done to the rivers and to Mobile Bay, the system remains remarkably productive. In large measure that's because of the incredible

amount of water that flows through it, the amount of forested land still remaining in the state, and the flushing effect of regular hurricanes, which sometimes suck up dozens of years' worth of pollution and whirl it out to sea. In essence, the system is in the shape it is in largely by virtue of luck, not through any human effort to protect it.

"The volume of water coming down is so great it flushes the system rapidly. The bay itself flushes in one or two days," said John Dindo of the Dauphin Island Sea Lab. "The contaminants coming down do not have a large enough residual time in the bay. That's very

Above: An osprey takes a morning bath in the Tensaw River in the lower Delta, the hulking cranes of the Alabama State Docks looming in the distance.

Inset: A flotilla of baby woodducks navigates the Blakeley River.

important. . . . it means the system can recover if we let it."

While some would argue that this natural resiliency means we don't need to do anything to protect the Delta, the old timers see it differently. Gazing back over multiple decades instead of a few years, they compare the rivers and Delta forty years back with the world they find today.

"I thank God for the Clean Water Act," said Robert Meaher. "I hate to think of where we'd be today without it."

Ladd said the problems related to DDT and chemical pollution have improved, but the dams have held back the natural spring floods and changed the way sediment moves through the system. Restore the natural flows, and the system might heal itself. Other than that, he said, "Pray. Pray for the Delta."

Meador said the state has got to find a way to control the mud flowing into the rivers and bay so the grasses can come back, "because those are our nurseries for all the fish and crabs." And John Valentine said a growing body of science indicates it is time to look at removing portions of the Causeway.

E. O. Wilson explored the Delta and Mobile Basin rivers as a Boy Scout in the 1940s. He watched its decline when he returned to Mobile in the subsequent decades. Returning to a beach he remembered from his youth for its clear water and grass beds, he instead found "murky water, no sea grasses, and a generally unsavory appearance." Still, he was buoyed by what he found when he began conducting research in the area in the last ten years.

"Let's hang on to this, what we have. Hunting and fishing won't harm it," Wilson said. "But what we should be doing is taking every measure we can to keep from harming it around the edges. The edges of the Delta are not the place for heavy industry. Alabama has plenty of places for industry, refineries, and timber operations. We need to make sure we do not continue practices that harm the water. In fact, we should reverse the damages we have caused so we have less turbidity, less silt and mud accumulation in critical areas. Alabama should set an example for other states to follow."

Wilson urged the state not to miss this chance. As Alabama's population grows, he said, "the pressures on its waters and forests will become more than the system can bear." That's what happened to the Chesapeake Bay estuary twenty years ago, when the oyster and crab harvests suddenly crashed under the pressure of massive development in the watershed.

"What will happen if we let it go? What will happen if we just do nothing?" Wilson's voice trailed off. "It would be quickly eroded and adulterated. I think we have a moral obligation not to let that happen."

Otherwise, ours will be the next generation reminscing about what the Delta once was and may never be again.

This black-necked stilt hunts against a backdrop that might have come from a Bergman movie. The creepy rust-colored ground was stained when the center of the island flooded. The groundsel trees died during the flood, rendering this stark landscape.

Buzbee's Fish Camp has been selling minnows and crickets at the mouth of Bay Minette Creek on the eastern edge of the Delta for a hundred years. The chemical factories surrounding Mobile are visible lining the Delta's western edge on the horizon. In between, nothing but swamp.

11

Protecting the Edges

The fate of the Mobile Basin will be decided along its edges. Whether the most diverse river system in North America survives into the next century intact—or loses ever more species to extinction—will be decided by how Alabama treats the forests and wetlands surrounding the water—from the banks of the smallest tributaries and largest rivers, on downstream to the shores of the Delta and Mobile Bay.

The track record to date is not a good one. Alabamians like to tout the fact that the state has more than seventy-seven thousand miles of rivers and streams. Left unsaid is how many of those waterway miles have been ruined. Sure, water still runs downstream in all of them, but how many function as they did two hundred years ago, before we began locking and damming and turning the waterways to our purposes? How many are still home to the incredibly rich assemblages of fish and mussels they have nurtured since creation?

You can find some answers simply by counting dams, and the acres of floodplain forests that were submerged, which measure in the millions of acres. You can find some answers in the EPA's annual 303(d) list of impaired waters, which shows the dozens of streams and river sections in the state that are classed as polluted by federal officials.

You can also find some answers in the ninety species lost to extinction, the hundred species now threatened with extinction, and the hundred more that biologists believe should be added to the federal list of creatures on the edge.

But more and more, it becomes apparent that we cannot fully understand what has been—or will be—lost because we haven't fully identified everything that lives in the state. Alabama is that rare place where scientists still find new species on a regular basis. We are not talking about the discovery of a barely visible bacteria species living in the mud on the bottom of a river. We are talking about the discovery of big, obvious plants and animals, such as new species of azalea, new species of witch hazel, new snails, insects, or even a salamander that has been hidden in plain sight for decades.

There are bright spots to be sure, such as the strong recovery of the Tulotoma snail, the first mollusk species in the nation to be removed from the federal endangered species list not because it had gone extinct but because its population had recovered. While it is still classed as threatened, the Tulatoma snail's recovery is tied directly to an agreement that required the steady release of water from Coosa River dams—the same dams responsible for more extinctions than any other factor in U.S. history. The snails now number in the millions.

Another bright spot is the rediscovery of the rusty gravedigger crawfish, a small, lipstick-red species that biologists with the U.S. Fish and Wildlife Service believed had gone extinct, largely because its only known home was in a small swamp sandwiched between an interstate, a mall, and a sewage treatment plant serving fifteen thousand homes.

This photo depicts the 2002 first sighting of a rusty gravedigger in twenty years, and proved the species was not extinct.

In this case, I played an unusual role—reporter turned resurrectionist. Having noticed that the only place the rusty gravedigger crawfish had ever been seen was in a swamp a few miles from my house, I decided to go looking for it. Finally, after three weeks, I captured a rusty gravedigger in the dark of night as it emerged from its burrow to hunt. It was the first one seen in twenty years. Sparked by the rediscovery, crayfish experts from other states came to Alabama to visit the site with me. The first day, they captured seven more rusty gravediggers. In the next week, they discovered the species in several other creeks in the area, with thriving populations. Since then, its range has been expanded almost to the Florida border. At the end of the day, the scientists concluded the species had been living in these other places forever. But no one had ever noticed.

And that, in a nutshell, is the conundrum facing Alabama. It is the most biologically diverse state by area, home to thousands of unique creatures and plants, but until very recently no one had ever noticed them, let alone thought to protect them or their habitat.

So what now? What can be done to protect the rare ecosystems in this richest of states? The answer is deceptively simple but will be devilishly difficult to accomplish.

To save what remains, you must protect the edges.

If you were to study Alabama's diversity, from carnivorous pitcher plants and rare orchids to the dozens of oak species or the nation's most diverse assemblage of aquatic creatures, there is one constant: Almost all of your research would occur within sight of water.

The water might not amount to much more than a puddle in a soggy, moss-covered pitcher plant bog or the small, ephemeral springs weeping from a hillside in a lush oak and magnolia forest, but wherever you are, there will be water. From the rocky rills at the top of the state to the cypress swamps and blackwater streams on the coast, that water is the reason Alabama has more than twice as many species per square mile as any other state on the continent.

When we talk about the edges, we are talking about the places where the land and water meet, along our rivers, our creeks, our bays, and our beaches. The zones straddling wet and dry are biological hotspots, providing the just right mix of water and habitat required by so many of our rare organisms. The problem is, this meeting zone is often where people want to be. We want to build our houses along the river to enjoy the scenery. Industry wants to site factories at the edge of the floodplain so their waste streams can be diluted in the river and washed downstream with minimal treatment. Ranchers want to use the creeks on their land to water cattle.

The challenge is to find a way to accommodate the needs and desires of people while protecting the water quality nature requires.

"We have done an incredible job of protecting about a hundred thousand acres of the Delta. That is just a remarkable thing," said Bill Finch, science advisor at the Mobile Botanical Gardens, discussing the purchase of the wettest, swampiest part of the Mobile Basin, the Delta lands subjected to the annual flooding. In truth, that land essentially protects itself from development simply because it is impossible to build on swamp mud that is underwater five months of the year.

"There was nobody spending a lot of money developing the Delta, so that land came cheap. It's great that it is now public access, and that's just a wonderful resource. But it is the edges that are so important to the Delta," Finch said. "The stuff that was harder to purchase because people had good views there, and they could build their houses on those areas . . . we have done a miserable job of protecting those areas, and that is the great challenge. I'll say it for the fifteenth time: If we don't protect those edges, we have lost the Delta. We have lost the Delta."

The same is true along the entire length of the Mobile Basin. If we don't protect the edges, we will lose this system and everything that makes it special.

THE NOTION OF CONSERVATION has had a bumpy road in Alabama, with the state's natural resources typically sold off to corporate interests headquartered in other states or other countries. Such was the story with the production of coal and

A snail—about the size of a marble—skates on a layer of water across the surface of a lotus pad.

steel and timber in the early part of the twentieth century. It was true again during the chemical boom of the 1950s through the 1990s, when international corporations built factories to manufacture paint, DDT, chlorine, fertilizers, pesticides, and paper. Most of the wealth generated left the state. Meanwhile, all of the pollution stayed here.

"It reached a point where we had factories with more than a thousand employees just discharging their sewage into the rivers. The Delta's wetlands were being used as dumps. It was bad. Things were being ruined," said Skipper Tonsmeire, who spent his working years in real estate and construction but has emerged as one of the most significant figures in Alabama's fledgling conservation movement. "There was a perception that because the Delta had always looked like it did and always been the same, that it's okay, we don't need to do anything up there. And then these five- and ten- and thirty-acre chunks were being developed, or mismanaged, or just absolutely

Effects of Humans: Above, a controlled burn in a protected Alabama pitcher plant bog reveals dozens of tossed glass bottles. Below, a newly hatched chick exhausted from kicking his way out of the shell lies on the graveled flat roof of a Piggly Wiggly. Lacking suitable nesting sites on Mobile Bay's heavily developed shorelines, least terns have taken to nesting atop commercial buildings.

destroyed. . . . It started to become apparent that things weren't okay, and we needed to do something."

Tonsmeire and a group of like-minded friends formed the Coastal Land Trust and purchased a hundred thousand acres in the Mobile-Tensaw Delta in 1986, a band that stretched clear across the giant swamp, from Mobile to Baldwin County. The group also purchased a large chunk of land on Fort Morgan, which became the Bon Secour National Wildlife Refuge, and a chunk in Weeks Bay, which became the Weeks Bay National Estuarine Research

Reserve. For a state that had always taken its rivers and bay for granted, the idea that private citizens would buy so much land with the simple goal of keeping it from being destroyed was a game changer. It served to inspire the creation of Forever Wild, the state's land trust program, which has protected about two hundred thousand acres in twenty years.

But even with Forever Wild, the state lags far behind most other states in protecting the land. Alabama has protected about 4 percent of its land, which is about the same percentage that was protected before the Forever Wild program was enacted. With thirty-three million acres of land in the state, the addition of two hundred thousand protected acres over twenty years represents slightly more than half a percent of Alabama's total land area—not nearly enough to make a difference in safeguarding delicate areas.

California has protected close to 47 percent of its land, while neighboring Florida has protected about 20 percent of the state. Accounting for the difference is as simple as looking at the amount of money each state dedicates to land preservation. In Florida, a real estate tax generates about $300 million annually to be used for conservation purchases. In Alabama, the state dedicates around $10 million a year from the hundreds of millions in revenue generated by natural gas wells in Mobile Bay.

The RESTORE Act funds expected from the BP oil spill will provide a huge windfall for the state that could be used to set aside and protect land along the coast and in the Mobile Basin. We must not miss this opportunity.

A BATTLE IS PLAYING OUT in the political arena regarding what role the federal government might play in protecting the rivers and creeks of the Mobile Basin. Championed by E. O. Wilson, the idea of creating a new national park caught the attention of former representative Jo Bonner of Mobile. In one of his last acts before resigning his seat, Bonner nominated an area that encompassed much of the Mobile Basin, at least the areas directly surrounding the rivers and larger streams, for consideration in the National Park Service Act of 2013, which authorized officials from the Department of the Interior to conduct a feasibility study.

The study became a central issue in the 2013 special election to fill Bonner's seat, which sits in a district so reliably red that Bonner twice ran unopposed by a Democrat. In the Republican runoff, a Tea Party challenger and the establishment candidate spent much of their energy promising to fight any effort by the feds to buy or protect land in the state. One of the first official acts by the eventual winner, Representative Bradley Byrne, was to torpedo the park study.

Lost in the political hubbub was this fact: the federal government—through the U.S. Army Corps of Engineers—already owns almost all of the lower Delta, having bought most of what Tonsmeire and the Coastal Land Trust saved in the 1980s.

Opponents of the park have now taken to calling Wilson a "half-Earther," a reference to his proposal to set aside half the earth for conservation in order to prevent a wave of mass extinctions. In Alabama, where a meager 4 percent of the land has been protected from development, it would take the purchase of perhaps another 3 or 4 percent of the state's land to protect the remaining rivers and valleys of the Mobile Basin forever.

The plants and creatures that inhabit these amazing rivers wouldn't care a whit whether it was the state or the feds who protected them, and it is important to remember that creating a park does not magically solve any of the problems facing the Mobile Basin.

No matter who owns the land in the Delta, more species will continue the long, slow slide into extinction due to the ongoing influence of the thirty major dams in the state. Runoff from clear-cut timber operations, construction sites, farms, and parking lots will still damage our rivers, smother our mussels, and destroy the seagrasses in the system. Lured by tax incentives, cheap labor, and lax environmental protections, new polluting industries will move to Alabama.

In order to protect the system going forward, those are all the challenges that must be addressed, but any efforts to do so are hamstrung by the state's anemic environmental protection efforts.

There has long been a school of thought among the staunchest conservatives that the best way to fight onerous environmental regulations is to cut the funding to the agencies tasked with enforcing them. That has come to pass in Alabama.

"We are the lowest funded program in the nation and have been consistently since about 2008," said Lance Lefleur, director of the Alabama Department of Environmental Management. "Since 2008, our general fund appropriation has been cut from $8 million down to $1 million. That's an 87 percent decline."

The state now provides about a sixtieth of the agency's funding; the rest comes from federal grants and permit fees. Due to budget cuts, the agency was forced to trim the staff to the bare minimum required by the EPA and sell off assets, such as the airplane required for aerial inspections. Lefleur pushed to double permit fees paid by industry to make up some of the loss.

"We have increased permit fees by approximately 80 percent in four years, but we are still at a net lower rate by about 60 percent," Lefleur said.

The agency writes about ten thousand air and water pollution permits a year. An industrial organization described Alabama as the most favorable state to obtain new permits,

Lefleur said, explaining that the state issues permits within 105 days of a request, compared to up to six years in California. He called the difference "a competitive advantage for Alabama," and in the last decade it has played a role in luring two new steel mills, two new car factories, and companies that make battleships and jumbo jets.

In a 2012 interview, Lefleur characterized the state's approach as creating a "bare bones environmental program . . . Industry wants the minimum amount of regulation; the state wants to keep government out of people's lives."

When I quoted Lefleur after that 2012 interview, he was vilified by the environmental community, who viewed him as selling out the state's rivers and wild places. But the truth is more complicated. In fact, since that interview, I've regarded Lefleur as a hero. Allow me to explain.

One afternoon, I received a call from Lefleur. "Hello, Ben. I want you to do something for me. I'm going to send you a link to an audio file of the most recent Environmental Management Commission meeting. Listen to my presentation. If you want to talk about anything, give me a call."

So I listened. And I was astounded at what I heard. The Management Commission is a board appointed by the governor to oversee the state's environmental agency. During his presentation, Lefleur told the board that his agency had been gutted, that the state legislature had wiped out the agency's budget. He warned it would be impossible to police the rivers and the industries that pollute them. He warned that the federal government would end up taking over the administration of the Clean Water Act. In short, he warned that environmental protection in Alabama was over. And then he called me, the last environment reporter then left in the state, and made sure I knew what he had said. When I wrote about his comments, I felt sure he would be fired. To my surprise, he is still in the same position heading the state environmental agency, trying to do what little he can to protect what's left. I came to see Lefleur not as a stooge carrying out the brutal desires of the most conservative state in the nation but as a last bastion of hope trying to make a scuttled ship sail into the wind.

"Someone has to take a balanced view," Lefleur said. "That's what our job is. It's not to stop industry, and it is not to let industry run roughshod over the environment. Those two things are not mutually exclusive, but they are conflicting."

Lefleur said the complexity and fragile nature of the Mobile Basin's rivers and wetlands makes for a difficult management task, especially so where industry and the natural systems butt together. But in his mind, the factories and industries regulated under ADEM permits are not the largest problem.

"Those non-point sources where we don't have a permit to control their runoff, that's the biggest area for improvement," Lefleur said, referring to pollution sources that are not governed by state environmental permits, such as heaps of garbage dumped in the woods, chemicals poured down the sewer, mud from tilled farm fields, or exhaust from cars. "Storm water is the biggest culprit. It carries gas, oil, and grease off parking lots. It carries silt from unprotected property next to waterways. You see it all the time on Dog River. D'Olive Bay, you can walk across it now, where you used to be able to fish it."

Lefleur said the non-point pollution problem is as small as half-acre construction sites and as large as the state's biggest cities. He noted that recently the agency was forced to ask a judge to "put Mobile into receivership in order to get their storm water system where it needed to be." For years, the city of Mobile failed to fund or monitor its storm water program as required by the Clean Water Act. The city was forced into compliance in 2013 after gaining national attention when, after every storm, I posted videos of trash-filled rivers, which

Papermill effluent is pumped into the Alabama River from one of the nation's largest pulp mills. In most states, factories must clean their wastewater of toxic compounds before it is dumped in a river. Alabama grants polluters a "mixing zone," provided contaminant levels are below EPA standards about a mile downriver. The foul-smelling wastes from this mill were easily detectable two miles downstream when this picture was taken.

went viral. The problem was so bad and so widely known that local fishermen targeted the floating trash mats in Mobile Bay after storms because fish congregated beneath them. After I had been putting up those videos for a few months, Sam Jones, then-mayor of Mobile, stopped by my desk in the newsroom.

"You are embarrassing the city with these stories about trash," he said. "You need to stop."

It was a stunning statement that summed up so many of the environmental problems we have in Alabama. The mayor wasn't looking to stop the massive amounts of garbage flowing from the city into the bay and the Gulf of Mexico. He was trying to stop me from telling people about it.

ONE OF THE BIGGEST problems with controlling runoff in Alabama comes down to the difference between rules and recommendations. In Alabama, the state uses recommendations to govern runoff from the timber industry, one of the most significant sources of the mud flooding into the rivers of the Mobile Basin. As a practical matter, forestry operations are obligated to keep mud from getting into the rivers. The state makes some recommendations for the best way to go about that, but Alabama does not enforce the hard and fast rules that protect the water in other states.

Trees are big business in Alabama. In fact, timber is Alabama's second largest industry, bringing in about $13 billion annually. Roughly twenty-three million acres are dedicated to growing timber in the state, two-thirds of all the land in Alabama. Only Oregon and Georgia have more acres of land farmed for lumber. With so many rivers and creeks crisscrossing the state, the impact of logging on the Mobile Basin is undeniable and massive. It has been that way for a century.

Most of the land in the state dedicated to timber is in

private hands, and most of those forests are allowed to grow naturally, as opposed to the densely planted loblolly pine plantations—which E. O. Wilson calls biological deserts—that cover about six million acres. In the natural forests, the land functions as habitat for wildlife and serves as a protective buffer for creeks or rivers running through it, until the owner decides to start removing trees. For decades, clear-cut logging, where every living thing is chopped down or crushed beneath heavy equipment, has been the preferred method of harvesting trees, mostly because it's the cheapest way to go. The big trees head to the lumber mill. The little trees are sent to the paper mill.

Drive along any country road in Alabama and chances are high you'll come across a forest that has been reduced to a moonscape of stumps, mud, and wood chips. One problem with such clear cuts is runoff. That's where the recommendations—or best management practices—for logging fall short. The recommendations include things like leaving a buffer zone around streams, planning logging roads so they avoid creeks, and using barriers to prevent mud from washing downhill when it rains.

But unlike other states, Alabama does not set a mandatory buffer zone around rivers and creeks. It simply recommends such a buffer, and then suggests cutting no more than half of the trees in the buffer zone.

In actual practice, forest crews in Alabama are legally allowed to cut trees right up to the water's edge, which involves heavy equipment in our most sensitive habitats. Visit a recent clear cut on a rainy day, and the problem with this approach is easy to see: massive amounts of mud flowing into the water when it rains, smothering creeks and river bottoms on its journey downstream to the Delta and Mobile Bay.

Contrast that with Oregon. There, officials insist on a mandatory buffer zone around streams. It is illegal to cut any tree or underbrush growing within twenty feet of the annual high-water mark of any stream in the state. That simple, concrete requirement serves several functions. The sliver of natural forest provides a barrier that traps mud flowing off cutover land. And it maintains a shady canopy over the stream, which can be critically important to aquatic creatures. The canopy helps maintain stable water temperatures and provides woody debris for the stream, generating the snags and logs that create habitat. The buffer also serves to keep heavy equipment away from the water's edge.

Considering that Alabama's forestry recommendations govern practices on two-thirds of the land in the state, it is not a stretch to say that enacting a law like Oregon's might do more to protect the rivers of the Mobile Basin than any other factor.

It seems a wicked twist of fate that the state with more aquatic creatures than any other—with more species per square mile than any other—is also the state that spends the least money to protect the environment.

It seems foolish that the state with the third biggest timber industry—also one of the rainiest states in the nation—would have no formal law protecting the banks of its rivers and streams from logging and runoff.

It seems a tragedy that more than two dozen dams choke the most diverse rivers in North America to provide just 5 percent of the electricity consumed in Alabama.

It seems like a failure of leadership that politicians would fight a federal effort to buy and protect more of the Mobile Basin without proposing an alternative plan for the state to step forward and take on this most important of tasks.

Alabama stands at the top of the list for extinctions in the continental U.S. The state has lost nearly twice as many species as the next state on the list. More creatures will vanish in the next decade. If you find yourself thinking that maybe the extinctions don't matter, you are not alone.

There is a common sentiment across the state that it does not matter if a creature as seemingly inconsequential as a snail goes extinct. Likewise, many imagine that the loss of a dozen acres—or a hundred acres—of wetlands won't harm the environment. Neither, they argue, will the pollution pumped from our factories into our rivers. But in Alabama, more than any other place, this is fiction.

Alabama is scarred and bruised from our rough treatment. The rivers and the Delta are under siege, battered by a thousand insults, great and small. Come with me and I will take you to a new railroad storage yard that sits on top of a salt marsh where I used to fish. Ride up the river with me and I will show you the outfall from a paper mill, where the fumes will choke us, and the river is lifeless and stained the color of Coca-Cola for more than a mile downstream. While we are there, we can stop at the dam where they caught the last Alabama sturgeon seen on Earth, netted as it tried to reach its ancient spawning grounds. That spot is just down the hill from a big clear cut that buried a lovely pebble-bottomed creek and all its inhabitants in mud.

But at each of those locations, we can walk a few feet and find something beautiful. That natural bounty is partly to blame for all we've wrought. It always seems we have more than enough nature, and that wrecking just a little bit for progress is a worthy trade. Perhaps. But even a child can predict how that approach will one day play out.

Think back through our history. Alabama was first recognized as a land of plenty, of natural wonders. Two hundred years ago, before coal and cotton and steel and civil rights, before we forgot that what was special about this place was

Four of Alabama's 204 snail species—from top: the smooth hornsnail; cylinder campeloma; Tulotoma snail; and Leptoxis plicata—highlight the state's world-class diversity. Seen up close, the intricate beauty of the snails testify to the value of every species, even something so lowly as a snail. (Photos courtesy Paul Johnson, Alabama Aquatic Biodiversity Center)

The delicate flowers of water stargrass invite pollinators. This is one of many plants gray ducks and mallards must have to forage in.

the landscape beneath our feet, before we let industry and neglect have their way with our woods and water for generations, Alabama was celebrated as a place to see things that existed nowhere else.

William Bartram, one of the nation's first great naturalists and botanists, cut his teeth in our forests. Charles Lyell, the father of geology, traveled the Mobile Basin because of the global reputation of our fossil beds. The undying curiosity of E. O. Wilson,

the preeminent scientist of his generation, was birthed in a Mobile-Tensaw Delta swamp. And it is here he returns in the twilight of his career to find new species and save what he calls one of the most special places left on Earth.

Remarkably, we still hold in our hands one of the nation's great wildernesses, despite the excesses of our forebears and the fresh insults added by our own generation.

The question, then, is what shall we do with what remains? It is ours to save. Or to lose. ❧

Index